12220976

STRATEGIES FOR
OFFICE AUTOMATION
Planning for Success in the Office of 1990

STRATEGIES FOR OFFICE AUTOMATION
Planning for Success in the Office of 1990

Walter A. Kleinschrod

The last of a four-part study on
"Managing the Office — 1990 and Beyond,"
sponsored in part by a grant from the Olsten Corporation

Administrative Management Society Foundation
2360 Maryland Road, Willow Grove, Pennsylvania 19090

Other books in the AMS Foundation's four-part
study on "Managing the Office — 1990 and Beyond":

Part One
The Office Revolution —
Strategies for Managing Tomorrow's Workforce

Part Two
The Office Environment —
Automation's Impact on Tomorrow's Workplace

Part Three
The Office Technologies —
Tomorrow's Tools for Automation Success

PREFACE AND ACKNOWLEDGEMENTS

Three excellent works in the series "Managing the Office: 1990 and Beyond" have preceded this fourth and final volume, and I feel privileged for having been asked to join such distinguished authors.

To their outstanding credentials as educator, human resources consultant, and office automation (OA) program director, I bring simply the resources of a journalist who has covered the office field for more than 30 years. Never doing, only observing, I nonetheless have formed judgments on what OA planning involves. Though my mission here was to bring back the story of what OA professionals have discovered about successful planning, certain perceptions of my own have inevitably shaped the work.

Thus, while I have many people to thank who helped me understand OA over time and with the present work in particular, the results are my own and I take full responsibility for them.

Among those I especially wish to thank are William Olsten, chairman and chief executive officer of the Olsten Corp., which partly funded this monograph series; the board of trustees of the Administrative Management Society (AMS) Foundation; and the AMS International Headquarters staff. I must also express appreciation to the many OA planners and implementors who took time from busy schedules to provide information for a survey upon which parts of this book are based.

In deeper research, I must thank Hal White of Monsanto Co. and Samantha Simonenko of Michigan Bell, who shared planning experiences and much OA knowledge with me.

Also: John Connell of the Office Technology Research Group; Dr. Raymond Panko, College of Business Administration, University of Hawaii; Sue Lyons-Gunn, Rubbermaid Inc.; the management of Geyer-McAllister Publications, Inc.; and William Olcott, editor of *Office Administration and Automation*.

Also: John Diebold, The Diebold Group; Randy Goldfield, Omni Group Ltd.; Lawrence Lynett, Lynett & Associates, Inc.; and N. Dean Meyer, N. Dean Meyer & Associates, Inc.

Finally, I give loving mention to my dear wife Pat, who endured long, lonely hours while I isolated myself with the cold, grey computer on which this book was written.

Walter A. Kleinschrod
Hempstead, New York
February, 1985

TABLE OF CONTENTS

INTRODUCTION

Plan carefully, plan thoroughly. The literature of office automation (OA) can't say it enough. Consultants preach it. Many a battle-scarred manager who has "brought OA in" knowingly confirms it.

Asked what he would do differently if he were to begin planning for OA again, a comptroller who implemented a $300,000 system in 30 days answers tersely: "Lengthen the time allotted."

A vice president of administration, whose more modest system preoccupied a three-member planning team for 10 months, replies: "Allow more time."

And yet a methods and procedures manager with a quarter-million-dollar annual budget for OA, whose system emerged from more than two years of painstaking planning, says the change he'd make, were he to do it again, would be to "speed the decision process." He is not alone among OA planners who would hasten the whole undertaking.

These conflicting attitudes on how best to approach OA are not the only ones to emerge from interviews, polls, and other research for this volume.

One survey, among subscribers to *Impact: Office Automation,* a newsletter edited specifically for office automation executives, even found differing ideas of what planners mean by "strategy" when they strategize on OA issues. For many, the big-picture view of the setting in which new systems will fit takes in their entire field or industry and its long-range competitive environment. But for many others, strategy remains internal, often reaching no further than a specific department or section.

On the question of considering the future while planning for the present, surveyed executives again divide on how far forward an OA vision should extend. Most say they look ahead three years or less. Others sketch in systems development plans which reach four years and more into the future.

Though an undercurrent of distrust for vendor promises and claims appears in many users' comments, nearly half the organizations reporting have asked vendors to take in-house surveys and make systems recommendations.

On the question of planning time — not how long it should take but how long it did take — the *Impact* readers could not have been more divided. From the onset of planning to the start of systems implementation, exactly as many respondents said less than a year had elapsed as said the planning took longer.

Even so, large areas of agreement emerge from this research. There is a consensus that top-management interest and support is vital to OA success. There is near unanimity that union activity or threats are not a problem during planning and implementation. And there is surprising agreement on the kinds of goals users seek from OA — surprising because, contrary to popular opinion, the elimination of jobs is the least important of the goals. Greater information and faster information flow are by far the paramount goals.

Perhaps the most valuable results of these probes into OA planning processes are not the tables of percentages, but the pages of practical advice given by OA management veterans "from the trenches." It's one thing to theorize about grand strategy and another to implement it, tactic by tactic, in a working, competing, distracted, imperfect business.

This book, last in the four-part series, "Managing the Office: 1990 and Beyond," focuses not on OA technology nor on OA's impact on the worker and workplace (the other parts did that), but on the process of getting ready to make OA happen.

This work represents a book of strategies, a book of tactics and a book of conflicting opinion on the best approaches for each. And yet, as these approaches all focus on putting technology *into* workplaces *for* workers, certain overlaps with the other three volumes cannot be avoided. Their separate subjects, like the departmentalized office, are seamless. They constitute administrative management, 1990 and beyond.

EXECUTIVE SUMMARY

Not long ago, most office change was episodic and selective — an improved system here, a new machine there. Today, as automation knits the office into a single interactive entity and extends its reach to workers and information hubs a building away and a world away, more than technology has changed; the quality of change itself has changed.

No longer do piecemeal adjustments suffice; planning now involves whole departments and companies seen in their entirety. No longer do choices come packaged as simple tradeoffs of costly but fast, slow but cheap, bigger but necessary. Today's choices mesh complex issues with long-term processes and "soft" immeasurable benefits.

No longer can that new, better office be drawn up as a single set of blueprints, with a place for everything and everything in its place. Achieving the automated office involves two, perhaps three, different sets of blueprints.

A Plan to Plan

Because the OA planning process, like the automated office which results from it, is so interactive and multidisciplined, it must be planned and managed. Here, then, is the first set of blueprints: a plan for planning. How elaborate the plan, how long the process (some say it never ends) are matters of conjecture, and later chapters deal with them in detail. There is a need to understand goals, set timetables, and assign people the task of defining how technology might help workers perform more effectively, and thus help the business reach its goals more certainly, within the allotted time. These elements are basic to the process, however quickly or painstakingly it is carried out.

From this effort comes the second set of blueprints — the basic plan of a proposed new system. This could fit on a few typewritten sheets along with one or two diagrams showing how text editors and personal computers might share a common database. Or it could flow elaborately from report to report, visual briefing to visual briefing, each discussing some aspect of the system and how it might grow in stages, ultimately linking mainframes to personal computers (PCs) to printers and other graphics tools over local- and wide-area networks, to which outside databases might also be tied.

The mechanics of presentation mean less, however, than the options and issues addressed in — or overlooked by — the proposals themselves.

How will this system affect the workforce? How favorably or fearfully do managers regard it? What are its implications for the clerical staff? These are crucial "people" issues explored by Dr. Harold T. Smith, C.A.M., in the first of this research series. Then, what will the impact be on the workplace; on office layout; on the need for, and capacity of, the premises to be recabled; and on the need for new lighting, to reduce harmful glare on screens? These issues of office environment are addressed by Wilbert O. Galitz in the second part of this series. And there is the technology itself. What hardware and software is out there, how reliable is it, and what can be expected soon? These are subjects of Alan Purchase's treatise, third in the series.

But what is the technology *for*? What is office automation (OA) expected to do? How well will OA work? How well can anyone know its impacts before it arrives? Answering questions like these is part of the planning process — the subject of this book, the last in the series.

One OA Plan or Two?

Trying to define what office automation "is for" shows how changeable the change process is. Hasn't OA always aimed at greater productivity, i.e. doing more with less? Certainly that was the intent of early OA systems and their predecessors, such as word processing. But now a more sophisticated rationale enters the planning, based on the idea that there are essentially two kinds of OA with two different purposes: (1) greater productivity; and (2) better performance.

The second purpose — the enhanced performance of executives, managers, and other office professionals — is where the bigger payoffs of OA can be found. In a fast-moving world, competitive survival will increasingly depend on the ability to use, create, and move information. Having the tools in place to support information flow throughout the office will matter more than having enormous centers to simply process masses of words and numbers, efficient as those centers may be.

Both types of OA systems are important, but they are different. This is why a third blueprint now often emerges from the OA planning process — one that involves automated systems designed for end users in addition to, and in contrast to, the automated systems for heavy processing of accounts, correspondence, and such other traditional bulk applications as word and data processing.

Of course, not everyone who plans for OA sees its applications quite that way. For many organizations, OA is still a mechanistic concept aimed only at speed-ups and cost cuts. And for some, automation means merely having a few word processors around. But lights do go on slowly as understanding deepens. Indeed, this progression of management awareness of the flexibility of OA may influence its development more strongly than any series of technological breakthroughs.

Before going further, let us define our own views. As the Administrative Management Society (AMS) Foundation explained to all it interviewed and surveyed for this book, OA means "the use or anticipated use of advanced office systems and information technology on a fairly widespread scale." This does not include the isolated use of a few word processors or personal computers.

As for other terminology, a number of critics and office watchers, notably John Connell of the Office Technology Research Group, favor *advanced office technology* over *office automation* on the grounds that OA sounds mechanistic. They argue that the main benefit of these systems is not the cutting of processing costs but the improvement of worker performance. Although we agree, we are not prepared to drop the term *OA*. It has earned its way through ample use (and misuse). We will use the terms *OA* and *advanced office systems* interchangeably throughout this work and sometimes add *infor-*

mation technology, too, for sake of variation.

Besides, there *is* a mechanistic, cost-cutting side to OA. It may not be the more important side, once minds have opened to see it, but it is still a viable use. Nearly half of the OA-using firms we surveyed (see Appendix A for detail) said cutting administrative costs ranked among their highest OA goals. However, more than 80 percent listed increased information accuracy and faster information flow as prime objectives. These are clearly the kinds of benefits that aid worker performance. Saving executives' time is another performance-related benefit, a goal two-thirds of the group rated high.

Whether planners come up with one OA systems blueprint or two may be a matter of how the executives perceive OA. Even with two systems, information must still flow between them as needed, and one technological structure will probably underlie both. And while each will be run separately (the end-user systems by their own department people and the central systems by MIS people with somewhat different responsibilities than most have today), some single management entity will likely oversee both. The one- or two-blueprint question may thus be a metaphoric toss-up, but the two-system model is worth managerial notice. It prefigures the shape of OA, 1990 and beyond.

The two-system model also suggests that the so-called information center may be only a temporary stop-off on the way to greater self-reliance among end users by the 1990s. Most organizations with information centers today have set them up as "aid and comfort" stations to which end users may go for advice, assistance, and training in the operation of office technology. They are also places where users can avail themselves of special equipment on a shared basis.

However, as a second generation of personal computers moves into the office, and as a new generation of users (many of whom had computers in school) sits at them alongside older co-workers who have also become adept at PC usage, the need for information centers may diminish — or their character could change. The programming users will need, and the help they may occasionally require, will come not from MIS or other in-house technicians, as is now the case, but from within their own departments. This is how many planners who look beyond the short-term see office automation evolving in the coming years. Not all planners, of course, have such extended vision. Nor is this to say that this is how OA will evolve in all situations, since specific needs may dictate variations from the norm.

Add On or Start Fresh?

Another way of looking at the one- or two-plan question is from the viewpoint of the automated systems presently in place (if any), and how existing tools may be used, or must be replaced, to move on to the kinds of OA systems now envisioned.

Planners starting from scratch, either because no OA systems exist or because what exists no longer measures up, could well produce one two-part blueprint or two separate-but-related ones simultaneously, and the distinction wouldn't matter. But planners seeking to build on existing systems inevitably produce more than one plan, and here the distinction matters because it introduces the concept of time.

Office automation is not built in a day, nor is it built to last forever. Yet, built well because its planning process was also built well, OA systems can endure and evolve as new needs and technologies present themselves. If, for example, in the build-on situation above, the earlier plan took into account the need for later expansion and even described that expansion in general terms, and if a strategy for phasing in later equipment helped determine the kinds of equipment first chosen, then any new plans could be refined and implemented more easily. And if these second plans take into account the next likely phases of expansion, so much the better later on.

In the early days of word processing, when the main idea was to reorganize secretarial work to make it more productive, the concept of two separate-but-related systems also took hold. On the one hand, there was typing, the principal task of the average traditional secretary. On the other hand, there were all the remaining tasks of the typical secretary: filing, running errands, and answering the phone. Word processing advocates said all these tasks could be handled better if the work was specialized. The typing could be done on those new technological marvels, word processors, in a WP center. The other tasks, collectively known as administrative support, could be assigned to other specialists in AS centers.

Looking back, we can see that this theory seldom worked out successfully. There was merit to massing all typing under professional supervision; where the change from the old secretarial order was carefully implemented and operations well managed, sizable productivity improvements resulted. But there was also human resistance and organizational foot-dragging plus certain technological surprises, such as the personal computer, which arrived claiming to be less expensive than the old word processors and generally more versatile.

The point here is not to rehash all the issues of the early days of word processing. The point, again, is to relate these developments to a time line of phased progress. Because the technology for automated typing was ready and waiting, the WP side of the WP/AS split was nearly always implemented first. Administrative support was harder to implement and not as many of its tools, like automatic calendaring and file management software, were available, as they are today. For many planners, it was "WP now, AS later."

For many today, it may still be a case of WP now, related forms of OA later; or heavily computerized, centrally run OA now, user-based, user-run OA later; or an investment in ample numbers of standalone personal computers for executive and professional use now, networking later.

While the options are varied, the success of any one of them depends on the time line and advance planning for future development.

Guessing and Gambling

Since one cannot foresee everything, forward planning may be no more than educated guessing and prudent gambling.

The "educated guessing" relates to office technology and the capabilities it may provide. But also, and more importantly, it involves predicting likely changes within one's own field, industry, and competitive sphere, and above all, within one's own organization.

The "prudent gambling" relates to the chosen system's probable ability to support (or at least coexist with) later system developments. It also involves looking risk in the eyes, and being willing to use today's technology for today's problems. It is foolish to wait for that problematical "something better," and not be overly concerned with tomorrow. Let us build later on what we have if we can, but if we can't, let us even now accept that fact and be ready in, say, five years, to switch to the new technology.

Looking at the big picture — a forward view involving time, technology, internal as well as external likelihoods, and plain unknowns — planning becomes the process of turning corporate strategy into operating tactics. That has a military ring, but the analogy is apt. In the end, all organizations either prevail against or succumb to competitive forces. Intellectual prowess built on timely, accurate information holds increasing advantage in the modern corporate struggle.

The "general staffs" who can equip their forces with the ordinance of informational superiority, and who plan with sufficient understanding of aims, resources, and inner strengths and weaknesses, stand the better chance of sharing the fruits of victory.

A View from the Trenches

Headquarters theory, however, does not always match life in the trenches.

The not-so-big plan. For example, the so-called overview taken by many OA planners lasts six months or less, from the time planning begins to the onset of systems implementation. This was the case with 20 percent of those surveyed for this book. The long view brought many executives no further than a particular division or section when planning OA; one-fifth forsook the chance to bring the entire organization into view, to say nothing of their field or industry.

A longer future view. Having come this distance, however, many OA planners do take a longer view of what comes next. Better than two out of five say their OA plans extend four years or more into the future.

Biggest difficulties. Significantly, the three obstacles OA planners encountered most on their way to implementation closely match major themes of the first three books in this series: problems with people, equipment, and organizational structure. One-fourth of the respondents called user anxieties toward OA a major problem. Lack of equipment standards proved equally troubling. The biggest headache, felt by one-third of the respondents, came from turf battles among departments when OA technology changed organizational relationships.

Inside aid. The AMS Foundation research confirmed what other studies have shown: Most OA planning is done in committee. Committees have been much disparaged as bodies which produce camels while trying to design horses. However, because OA is a complex, multidisciplined undertaking, two-thirds of the organizations in this survey took the camel risk to gain the benefits of an opposing idea: "No one of us is as smart as all of us."

Yet, 33 percent of the organizations opted to place responsibility into the hands of "one of us." From their titles, OA planners, alone or in groups, appear to be an eclectic lot. Management generalists, systems technicians, and specialists from other disciplines mingle freely on planning assignments. (See Appendix B for all titles from the 45 surveyed organizations.)

Outside aid. Most OA planners are astute enough to occasionally leave the trenches and seek outside knowledge. Six out of 10 visit other organizations' automated offices; 87 percent attend seminars. Less favored are memberships in user groups (38 percent) and the use of outside consultants (35 percent). However, many said they would use consultants were they to do it all again.

Top management support. With benefit of hindsight, a number of planners said they would obtain, maintain, and demand stronger top management support. Even so, top management support scored well. More than one-third of the planners said senior people have been personally and actively involved throughout. Another 45 percent said that although early active involvement may have ebbed, top management continues to show interest by having periodic briefings.

Findings from this and other studies will be described more fully in later chapters. In general, it appears that those organizations surveyed for this book will meet their OA objectives.

Prevailing Themes

Circumstances differ; emergencies occur. Not everyone has the luxury of seemingly limitless time in which to plan and refine OA. Where administrative fires must be put out, as in cases described in Chapters 1 and 5, the sooner a do-it-right OA system can be set in motion, the better.

Still, a prevailing theme runs through most planners' views and experiences: Allow ample time for planning. Plan thoroughly, get as much input as you can. From the first moments of a perceived need for OA to the setting of early goals, and from technological assessments to the steps of implementation and the laying of paths to the future, see the process on as long a time-line as possible.

At every stage, a multidimensional body of needs, ideas, and realities advances along that time line. Issues of work, workers, and workplace interact within that moving matrix, taking shape as they go. What begins as abstract strategy becomes more visible in research findings, decision schedules, implementation plans, and finally, tangible office systems. It is a process of creation in an information ecology. Then, as in all ecologies, the process recycles. The useful grows and strengthens, and the useless eventually withers.

But unlike nature's ecology, where change takes care of itself, the changes required of office environments are for management to make. Advanced office systems are tools of change — and changing tools. Their purpose may seem to change, but it only unfolds as management's awareness grows. That purpose is as basic and inexorable as nature's: survival, growth, and fulfillment for the organization and its many living members.

Chapter One

OA PLANNING IN ACTION

Let's get right to specifics: How do the planners plan? From the following real-life vignettes, the answer is obvious: very differently.

Most case histories in the literature of office automation (OA) focus on benefits gained. The experiences recounted here are mainly success stories, too, but the focus is different. We examine the steps leading up to the introduction of new technology. Where possible, we address management lessons learned after the technology had been in use for a while. Later chapters discuss these steps and lessons from the broader perspective of future systems planning.

Whatever the particulars of individual cases, all OA systems begin with an idea or perception that the use, or greater use, of advanced office technology might benefit the organization. It is at some point after this perceived need has been felt that we pick up these individual stories.

Pilot Testing

In December, 1982, Warner-Lambert Co. decided to test the feasibility of OA as a productivity tool for managers and other professionals at its headquarters in Morris Plains, New Jersey. This would not be Warner-Lambert's first encounter with advanced office technology. The big pharmaceutical company already had a sizable department called Management Information Systems - Office Information Systems (MIS-OIS), essentially responsible for large-scale computer operations. MIS-OIS, especially the OIS side, would now look to individual computer use. It would set up a pilot test of a Xerox Ethernet local area network (LAN) which, if successful, could eventually serve executives throughout the organization. Initially, however, to keep matters under control, OIS would begin with its own department.

By April, 1983, a Xerox systems analyst had taken up temporary residence at Warner-Lambert headquarters to work with OIS executives. Part of the planning group's mission was to locate specific test sites and develop applications and training programs for OIS personnel. From this came the design of the pilot LAN. It would link eight Xerox 860 Information Processing System terminals and, later, six Star workstations, along with "server" units controlling files, printers, and communications. The purpose was to determine the feasibility of information transfer among all devices and "the extent to which the productivity of managers and other professionals could be improved."

The test proved so successful that by early 1985 the LAN had spread to four other departments. Seven more were in the process of being added. Benefits — while not the main subject here — were outstanding enough to earn the firm an Office Automation Award from *Office Administration and Automation* magazine. In brief, OIS, from initial tests to later established use of its system, had realized a 32 percent savings in executives' project-work time, a 40 percent savings in input time, a 52 percent increase in output, and a 31 percent decrease in overtime — a big factor in improving the quality of worklife.

The pilot approach, while often criticized for being less test than model for future systems expansion (see Chapter Nine), did alert Warner-Lambert planners to certain problems for which they could make mid-course corrections.

Environmentally, they discovered that equipment generated considerable heat within OIS office cubicles. More space had to be allotted for proper ventilation. Also, an early shared-device concept did not work out; planners found it better to give each analyst his or her own keyboard.

Technologically, OIS learned that workstations cannot be spaced too closely along the network "bus," or main artery. Installers were compelled to loop lengths of cable to maintain at least eight feet of separation between the machines.

Procedurally, OA led to duplicate files and a proliferation of electronic mail. Both excesses were later corrected. Productivity increases created management bottlenecks. For example, under the old system, MIS-OIS manager Joseph McGrath reviewed all jobs. Now they were arriving at his desk so fast he had to delegate some of this work.

On the people side, planners discovered a disquieting "loss of human contact." Also, as *Office Administration and Automation* reported, "The devices exercised a narcotic effect on employees." Workers were less willing to be interrupted. Happily, new value-added work filled the time the system saved.

These lessons are being heeded not only at company headquarters as OA expansion continues, but also at Warner-Lambert Canada, where a smaller Ethernet has been implemented, and at Warner-Lambert Ltd. in the United Kingdom, where a pilot program has been under way.

Perhaps the most important insight gained at Warner-Lambert is that you can't automate everything. "In fact," says McGrath, "we decided *not* to automate certain tasks."

Aiding him in the project were MIS-OIS supervisor Richard Cella, office services supervisor Wanda Sanford, senior office systems analysts Lisette Curry and Regina DeFiore, and analysts Phyllis Baldwin and Warren Schaub.

Interdepartmental Committees

In February, 1983, a large Midwest utility set up an interdepartmental steering group to "direct and control" OA through three stages: acquiring it, introducing it, and using it.

The group is elaborate in structure and far-reaching in its mission. Even the documents which set it up are sizable (parts appear in Appendix B).

Representatives from seven company departments or "segments" share the unit's work; four members are from finance, four from marketing, two each from personnel relations and support services, and one each from law, network services, and regulatory and external affairs. They form the Advanced Office Systems Steering Group (AOSSG).

Within AOSSG lies the Advanced Office Systems Coordinating Team (AOSCT). AOSSG "provides centralized direction" for OA company-wide, according to its charter. AOSCT's function is to "review and concur" in all future OA and word processing system installations. The founding document further decreed that AOSSG would be jointly chaired by support services and an eighth, centrally responsible department, the Information Systems Organization (ISO).

"What it all means," said a steering-group member, "is that we were stretching beyond word processing. We needed more staff, and we needed to focus on future needs. We had an MIS department which originally handled data processing and managed our utility service network. That department became ISO. Meanwhile, Support Services gave staff assistance and guidance to word processing and other office functions. We saw the time had come to try and put all this together."

A visiting interviewer asked if AOSSG had any overall corporate strategy to guide it. No, the group member replied, but the eight-person AOSCT unit was attempting to develop something close to a corporate strategy. There was as yet no top management input, so calling it corporate strategy was going too far. But a basic plan was evolving, most likely for furthering OA through existing Burroughs and IBM mainframes.

AOSSG's charter does give some direction. It defines advanced office systems (AOS), as "a broad term which includes most of the computing and communications technologies that directly support managerial and administrative personnel in the office." According to the paper, AOS positions the manager as "focal point" of office systems, and includes, but is not limited to the following:

- WP and other forms of electronic document and message preparation;
- electronic mail and other forms of document and message communication;
- calculator capabilities, automatic calendaring, project and time scheduling, and other managerial work aids;
- electronic document storage and retrieval, locally or remotely, through time sharing and other means, and the ability to reproduce accessed documents on such devices as printers and intelligent copiers;
- query functions, database management, and other forms of localized, "personal" electronic information processing; and
- all other applications using advanced technology to "improve managerial productivity or administrative support in the office environment."

Nonetheless, into its second year, AOSSG found itself being guided by top-management directives. These were not aimed solely at AOSSG but at departments company-wide. The orders said, in effect, "Cut budgets." As a result, planning in both AOSSG and AOSCT was shifting toward the cost-cutting forms of OA. The executive forms, with their highly supportive but financially less-direct payoffs, may have to wait a while.

This large-scale, in-house committee approach seems to fit the company style. "We're not big joiners," the OA planner replied when asked if the steering group sought help from user associations or outside consultants. A few support services supervisors do belong to the Association of Information Systems Professionals (AISP) and the Wang Users Group. The setting up of a well-defined collegial committee to develop office systems "fell right in line with long-established methodology."

A Well-Mapped Journey

"Automation really gives you the time to do the things you should have been doing right along," declares Robert Fowler, Jr., president of Rubbermaid, Inc., a major business supply and housewares manufacturing firm based in Wooster, Ohio. Automation does so by providing executives with greater access to information, says Fowler.

Acting on these convictions, Fowler, in December, 1981, authorized a comprehensive study of communication needs throughout the company. Completed by March, 1982, the study spelled out the equipment and training requirements that could satisfy the needs identified. Once management gave the go-ahead to automate, most important was the follow-through which converted the study from a mere document into a working OA system, phase by careful phase.

According to OA manager Sue Lyons-Gunn, who directed the planning and implementation, each phase was plotted along a strategic five-year time line. Every addition was accorded a reasonable amount of time to be physically set up, tied in with what already existed, and properly run by the people it was designed to assist. The personnel had, in effect, an explicit calendar of progress and a detailed map showing where they were at each stage of this OA journey.

The plan also had the advantage of strong top-management backing. Fowler and other senior executives recognized that the introduction of OA could buck long-held department traditions. They made sure that users knew management was giving these proposals its complete support. Whatever serious resistance there might have been was quickly overcome.

Because Rubbermaid operations are spread across the United States and into Canada and West Germany, the basic strategy was to automate specific functions at particular locations, integrate the functions locally, and develop the necessary links among divisions and plants. Each step brought the firm closer to its goal of an integrated information network.

At in-house orientation sessions, Lyons-Gunn often presented the five-year plan in the form of a pyramid. It rested on a commonly used base for OA development — word processing — along with such related functions as central dictation and facsimile.

At the next higher level, advanced communications capabilities would be added, connecting the "foundation" systems to out-of-house photocomposition equipment and optical-scanning devices. Then would come electronic filing and other advanced functions up to the integrated-network apex (see Figure 1-1 for detail).

The OA effort first took hold in Wooster's executive area, then spread to the Home Products Division, also in Wooster, and to Commercial Products, located in Cleburne, Texas, and Winchester, Virginia. Each division, of course, had its own needs, but Lyons-Gunn and her staff placed great emphasis on the ability to interconnect widely dispersed sites.

In July, 1982, right on schedule, selection began among WP vendors. Wang won out, and by October, installation of its OIS text editors got under way. By November 15, Jan Dudsak, supervisor of

---

Figure 1-1
Rubbermaid's OA Pyramid Plan

---

7. Integrated information networks and systems;
6. Multifunctional management workstations;
5. Multifunctional administrative support workstations;
4. Intelligent copiers and laser printers;
3. Electronic filing and shared databases;
2. Advanced communications, optical character recognition (OCR), and photocomposition;
1. WP, dictation, facsimile systems, and space planning.

OA training, was conducting the first classes in equipment usage. Meanwhile, other members of the planning team, notably Home Products OA analyst Judy Mack and supervisor of MIS training Lynn Bauer, were preparing for the next scheduled events.

As in most large organizations, OA at Rubbermaid eventually became a multi-vendor undertaking. By mid 1983 the organization had reached Level 4 on its pyramid, and by then 19 IBM PCs were in executive hands. Four-Phase, Hewlett-Packard, and Honeywell equipment in various parts of the organization was also awaiting interconnection.

The sudden arrival of PCs around this time was not foreseen in the original five-year timetable, as it was not in many firms' OA plans. Rather than predetermine their role, the way so much else had been predetermined, Rubbermaid this time took a different tack. It allowed managers who saw a need for PC applications to freely experiment with them, and to report their findings on special application-profile forms. Essentially, the users were being asked to assess the value of PCs in their work, and to suggest how, when, and where they might be integrated into the developing network.

Rubbermaid's OA effort has been noteworthy for its ability to identify needs and to provide systems which address them in a well-ordered progressive plan and in well-defined time frames. It has laid out and followed a clear itinerary, yet has also reacted flexibly to surprises along the way.

Big Firm: A Two-Person Study

A multinational firm with four of its companies based at St. Louis headquarters, the Monsanto organization has had a strong commitment to information technology for years.

In the course of those years, management of that technology has changed markedly. Put simply, it has 1) decentralized; 2) spread from conventional DP and MIS operations into the end-user office; and 3) made heavy use of the information-center concept to help end users deal with the technology properly.

Interestingly, many of these changes, potentially affecting thousands of headquarters workers, were the result of pilot studies directed by only two people.

According to central MIS manager Hal White, all this began to evolve in the late 1970s and early 1980s. Word processing was just getting started. The information-center idea was taking hold, but mainly as a resource for "techies" involved with MIS development. Most typing and mail distribution was still manual. Even so, ideas of what OA could do for business in an "office of the future" were very much in the air.

The MIS department set up a small OA group to pilot-test ways to go. "It was essentially a two-person show consisting of Russel Sprague, whose background was multifunctional and not DP oriented, and Wanda Gant, who had been the secretarial supervisor who ran our first word processing," says White. Three years after their effort began, a major event spurred intensive action. Monsanto, corporately, had decided to decentralize. What had been a centralized MIS operation now had no choice but to decentralize widely. With so much new planning needed anyway, it was an opportune time to really push OA.

Sprague and Gant started evaluating companies like Hewlett-Packard, IBM, and Wang — 20 vendors in all — and began to develop OA strategies that built up from WP and outward from MIS, toward individual computer usage.

As OA evolved over four years of trial and acceptance, information centers also spread to help the new end users. When interviewed, White said Monsanto had 15 such centers divided into four groups: office support, user support, data support, and executive support. Despite the distinctions, all were basically office centers, according to White. There was also his own, 16th center — "an information center for all the other centers."

Today, following Gant's recommendations, IBM Displaywriters and 5520 terminals for WP tie into the Monsanto companies' mainframes. Justification for each new system proceeds office by office, usually at the recommendation of the MIS manager responsible for that section; there is no grand plan.

Nor is there any overriding corporate strategy, in the highest business-driving sense of that term, to guide OA futures. But there are, White explained, certain corporate-wide OA strategies being developed, which is a different matter. For example, there are now directions established for electronic mail. There are corporate policies limiting major systems purchases to three vendors — Digital Equipment Corp., Hewlett-Packard, and IBM — with each of the Monsanto companies using at least two.

More definitive office plans continue to be formulated, still by that two-person team. Its recommendations, of course, must pass MIS muster. A worldwide MIS Council meets biweekly to approve (or reject) major plans.

Having seen such technological change in so large an organization, White was asked what "management maxims" other OA planners might heed. "People are the biggest concern," he says without hesitation. "You must plan on educating them to understand the technology and to commit to its use. Then, be careful in laying the groundwork for systems integration. We all still need vendor help on that. While many products perform well individually, overall they are not well integrated," says White.

Because of this, Monsanto, in some ways, maintains a wait-and-see position, adds White; it is not moving ahead with OA development as fast as it might. Even so, White sees a big difference in the way company data will be treated in 10 years: "Today, it's handled one on one. Tomorrow, there will be more formal management control of information. Information is an asset, and that has to be realized. A department like marketing may be the one to sell that."

"Keep it Simple"

To test an OA prototype, the Fort Collins, Colorado, city government added equipment to its existing police department DP system. A year later, having developed a set of new requirements as a result of that test, it called upon the mainframe vendor to help carry out the objectives. These included tying PCs and WP terminals onto a private branch exchange (PBX) network citywide. To everyone's surprise, the vendor could not fulfill the requirements.

Marsiea Yates, one-half of a two-person OA planning team, wasn't daunted. "We surveyed the market for new technology," says Yates; she found it. She refused to get locked into proffered turnkey "solutions" that did not fully provide what was needed. "Now, because we planned thoroughly, we've had few surprises," she says.

Despite this thoroughness, Yates' advice to others is, "keep it simple." She refers not so much to the technology as to the management of the project. Recalling the experience with vendors that might have led others to compromise, she says, "Evaluate the needs of your users based on their own input. Don't let overwhelming 'options' sway you from your planned course in meeting those needs."

Prior to joining a newly formed OA department as manager of integrated office systems, Yates had been records manager. Her colleague in the year-long OA planning and implementation effort, formerly assistant finance director, became director of information and communications services. "We were given free rein to make OA happen," Yates says. Even so, higher authority within the city administration was actively involved in the early stages, and continued showing interest through periodic briefings.

The planners worked within an initial two-year budget of $600,000. Operations now run at about $250,000 annually, with various departments responsible for securing additional equipment.

"Use Experienced People"

As OA projects go, it did not seem overly taxing. At a packing company in central Canada, a small systems department would be added to an existing DP department to develop PC applications. Some new equipment would be needed on which to create and test programs.

With benefit of hindsight, an office manager who took part in the four-month planning effort said that if he had to do it again, he would bring in more expert help.

"Have experienced people on site during the planning and implementation stages to oversee progress and assist in eliminating problems," he advises. Then:

- "Know your present systems thoroughly."
- "Visit a user site and get their feedback."
- "Document your plans and schedules."
- "Make sure your chairperson is experienced or at least has been thoroughly briefed."

"Operate on Two Levels"

A plan to "loosely couple" word processors and other OA devices to a mainframe has been under discussion and revision at a Southern service-merchandise firm since April, 1983. To date, nothing has been implemented. A problem may be that *everything* has been loosely coupled, including the style of the planning.

"I became involved midway through, but my observation is that OA should have been organized under MIS and integrated with PC and networking plans from the start," said a member of the firm's planning team, who asked not to be identified. Organization was left unclear, however. As a result, departmental turf battles have erupted.

Indeed, service on the planning committee itself became a some-time effort. Presently consisting of five administrative or MIS executives, with an MIS vice president as chair, the group has had different (and differing) members over time. The above-quoted source ranked "inability of planners to agree" as a problem almost as bad as the turf fights.

His advice to others: "Operate on both of two levels. One — get top management support by identifying quick-payback, obvious-benefit projects to gain early credibility. Two — use a committee with MIS commitment to define long-range issues and clarify the OA turf."

It's said there are two things the public should never see being made: laws and sausages. Cynics might add OA plans. But no, it is better to know what goes into such plans and what one can expect to get out of them, in order to manage the process with reasonable understanding and with benefit of others' experience.

For all the differences of approach and situation which these experiences portray, they also reflect patterns useful in guiding planners toward OA success. We examine those patterns in the following chapters.

Chapter Two
BUT WHAT IS OA FOR?

A study of office automation policies conducted by Omni Group Ltd., a New York-based consulting firm, in late 1983 hinted at something that many managers felt was true but were afraid to say out loud: OA was no longer the cost-cutting, staff-reducing marvel it was proclaimed to be.

What it was becoming was marvelous enough: a powerful means of enabling workers to perform better. And in offices, where most work involves information, systems which allow people to use information more effectively are central tools of the trade. This is OA's greater claim.

As shown by the AMS Foundation survey of OA-using organizations conducted for this book (see Appendix A for detail), better performance is OA planners' current, larger goal.

More than 80 percent of the firms which have implemented or are planning to implement OA cite *increased information accuracy* and *faster information flow* as their chief OA aims, according to our study (see Table 2-1). In addition, 64 percent rank *saving of executives' time among their most important goals*.

In contrast, only 11 percent mention *reduced personnel* as a high priority (see Table 2-3), with 41 percent rating it low in importance (see Table 2-2).

Two years earlier, Omni reported that half of the large *Fortune* 500 class of OA users ranked staff reductions as an important OA goal. Among small- and medium-sized organizations, however, only one-

Table 2-1
Office Automators' Most Important Goals

OA Goals	Percent of Respondents Rating Goals as High in Importance
Increase information accuracy	82
Speed information flow	81
Save executives' time	64
Cut administrative costs	48
Provide new services	41
Reduce personnel	11

Note: Percentages are composites of 4 and 5 ratings on a scale of 0 to 5. See Table 2-3 for a finer breakdown of ratings. Forty-three percent of respondents also cited various other OA goals, most of which rated high in importance; they are discussed more fully in the text.

Table 2-2
Office Automators' Least Important Goals

OA Goals	Percent of Respondents Rating Goals as Low in Importance
Reduce personnel	41
Cut administrative costs	8
Save executives' time	7
Provide new services	5
Increase information accuracy	4
Speed information flow	2

Note: Percentages are composites of 1 and 0 ratings on a scale of 0 to 5. Chart is thus an approximate statistical mirror image of Table 2-1. See Table 2-3 for a finer breakdown of ratings.

Table 2-3
How Office Automators Rate Six Major Goals

| | (% of Respondents Rating Goals) | | | | | |
| | Low | | IMPORTANCE | | | High |
Rating:	0	1	2	3	4	5
Goals:						
Cut administrative costs	2	6	13	31	22	26
Reduce personnel...............	23	18	18	30	9	2
Save executives' time	2	5	11	18	32	32
Increase information accuracy	2	2	7	7	30	52
Speed information flow	2	—	3	14	19	62
Provide new services............	—	5	21	33	20	21

Note: Figures show percentage of respondents giving particular ratings to various goals. In a fill-in question 43 percent of respondents cited additional goals, described more fully in text.

third rated staff reductions as extremely or very important. It was a sign that attitudes were changing. The era of staff cuts was perhaps playing itself out.

After all, by the mid 1980s, that big productivity gainer, word processing, had had a decade or more to wring the fat out of such operations as routine correspondence and similar document work. In addition, data processing had converted invoicing, inventory control, and the other labor-intensive tasks of mass accountancy into production lines of almost factory-like proportions.

It was time to put the text-generating powers of WP and the computing abilities of DP into the hands of the myriad executives, professionals, and other information users who work in offices, but who aren't necessarily there for "office work." It was also time to expand the users' reach through new forms of telecommunications and expand their knowledge through access to databases.

Seen broadly, the history of OA has been as much a series of managerial mind openings as of technological breakthroughs. First, there was the awakening to automation's mechanistic abilities to cut administrative overhead. Now, there is the understanding of OA's role as an information tool for competitive advantage through better performance among individuals — gains not as easily measured but enormous in cumulative impact. And perhaps just dawning is the realization that as technology changes work habits, certain overhead costs fall away, even though this is no longer the main reason for OA.

Put bluntly, the fact that reducing personnel is not a major OA goal for organizations like those in the Omni and AMS Foundation survey groups may simply mean that those reductions have already taken place, at least for the majority. These are not neophyte groups; they exhibit maturity toward OA, and they are moving forward. Then, too, for some, calling staff reductions a minor goal may simply mean that no one gets fired (not that staffs won't shrink through attrition).

How OA Alters Work

John Connell, executive director of the Office Technology Research Group, a well-known OA user forum, cites the following example of how technology changes work habits and even organizational relationships.

"Take a fellow in marketing who has a problem," Connell postulates. "He gets on his electronic message system and says, 'I've got an important customer complaining about this product malfunction. Who can help me?'

"He outlines the problem, and it's broadcast over the system. It goes to a number of people — people the marketing man doesn't even know. But someone on the engineering side of manufacturing sends a message back: 'Have you tried such and such?' And it solves the problem.

"Now, that communication link never existed in the past. There was no relationship between marketing and engineering," says Connell. But from here on out, when that marketing fellow has a problem, he'll remember this incident. He's got a new resource to turn to. An impact of this technology is that new relationships emerge."

Connell sums up: "There are only two reasons for investing in technology when you get right down to it. One is to streamline processes, and that's been the role of data and word processing. But the other is to improve your ability to compete. That's what that fellow in marketing was really doing. He was finding a new and faster way to help his customer, and in one small way he was putting his company in a better competitive position."

Staff Cuts — The Second Round

More can be said about how OA changes working relationships. As seen by the previous example, electronic mail, being so electronically direct, bypasses traditional methods of interoffice communication. There is no need now for memo-dictated, memo-typed, memo-carried out-box to in-box, reply-dictated, reply-typed, and reply-carried out-box to in-box, to facilitate the kind of interchange described in Connell's example. Layers of administrative personnel could well be cut as a result of these new capabilities, even though, again, an opportunity to cut staffs was not the reason the electronic message system was installed.

We come close here to a philosophical argument as to whether layers of personnel — including managers — are eliminated because OA technology allows the remaining people to work more productively, or whether people are eliminated because the technology spotlights areas where they were not needed in the first place. Connell, for one, strongly believes the latter. "Sure you can eliminate a layer of management, but did technology do that? Of course it didn't do that. What technology did was open peoples' eyes to that fact.

"Technology changes many things, *including* perceptions," Connell remarked in an interview. "It's like when a recession comes along and suddenly companies wipe out all kinds of staffs. How come? Well, the recession changed their perception of the need for these peoples' services. They found they really didn't need all of them."

Whatever the dynamics of change, and whatever the motivations of management in fostering it, OA clearly has an impact on people. Companies can *say* they have no intention of eliminating jobs, but what are they to do when it's plain that certain jobs have become superfluous? No wonder almost half of the firms in the AMS Foundation's survey rank *user anxieties toward OA* as a bigger-than-average problem (see Table 10-3, Chapter 10).

Cutting Administrative Costs

While it is encouraging to see greater emphasis being placed on the creative, work-enhancing uses of OA, it would be foolish to belittle its raw, cost-cutting powers as suddenly unimportant. They are important, as 48 percent of the surveyed OA executives were quick to acknowledge.

For all the publicity OA has received, it must be remembered that a majority of organizations have yet to introduce it in any meaningful way. While most *Fortune* 500 firms do employ advanced office technology, the Omni study showed only a third of medium-sized organizations and less than a quarter of small (under 100-employee) firms as having even a policy or strategy regarding OA. When non-using firms do begin to automate, it is only natural that their first targets are those basic processes involving documents and records which yield so readily to WP and DP systems.

It is significant, then, that more than half of the respondents to the AMS Foundation survey said the systems they were planning or implementing were start-ups. They were not being added to existing systems. While some of these could be second starts, it is clear from replies that for many organizations, this was their first OA system. Significantly, 78 percent of this group gave high ratings of 3, 4, or 5 to *cutting administrative expenses,* 5 being of highest importance.

Providing New Services

In the language of OA in-groups, OA's two main benefits — productivity and performance — are often referred to as cost-displacement or cost-containment benefits (depending on whether you want to cut costs or simply hold them where they are), and value-added benefits (doing new and better things that you didn't have time for before).

Value-added benefits are clearly what four out of 10 planners have in mind who say *providing new services* ranks among their highest OA goals. The nature of these services are as varied as the pursuits of business.

One organization seeks to implement a fund-raising competition. Another, a Calgary dairy, seeks a way to coordinate reports. And a small Michigan insurance company will soon communicate text — a commonplace function, perhaps, but an important step forward if you never have done it before.

Adding value has a nice business ring to it. Surrendering value — meaning paying for all of these systems — is something else. Are they worth it? And how can you tell? That vexing issue of OA cost-justification is what we look at in the next chapter.

Chapter Three

JUSTIFYING THE COSTS

Pity the planner trying to sell a skeptical, bottom-line type of management on the very real "but just a little soft" advantages of office automation. Describing the technology is difficult enough. But attempting to prove the near-unprovable with cloudy-bright terms like "enhanced performance" and "value-added results" could sink the project, if not the planner, on the spot.

Yet something close to that is what many OA executives face as they try to justify the not-inconsiderable sums budgeted or projected for office technology.

Office automation budgets for the 32 firms willing to provide the AMS Foundation with this information average $1,268,937 annually. Two budgets, one for a major pharmaceutical firm, the other a provincial government insurance system, exceed $5,000,000. Nine executives also provided OA budget data for specific departments, which averaged $297,222.

Distinctions between the two types of OA systems already described at length are never sharper than when one tries to cost-justify them. Some yardsticks can be applied to "measurable" work like word processing — lines per day, pages per day, documents per week — and it is not difficult to compare "the way it was" with "the way it is," and what was spent with what was saved. Indeed, a few norms, to be used with caution, are published from time to time, based on accumulations of many such cases.

But what of the very random, episodic, decisional types of work performed by managers and other office-based professionals? How can one quantify them? "With difficulty," all will admit, yet some valiant attempts have been made.

Performance Accounting

One early and widely quoted effort, the 1980 *Booz Allen Study of Managerial/Professional Productivity,* tracked the daily office activities of 299 executives. From the more than 90,000 separate time samples taken, the report's author, Harvey Poppel, found that the executives spent two-thirds of their time communicating. Poppel concluded that if the executives had been equipped with proper OA tools and became adept at using them, they could soon accomplish these same tasks with a 15 percent savings of time.

Significantly, 15 percent amounts to more than an hour a day — time to address new work which otherwise could not have been done — a value-added benefit. On the other hand, the company could pool the "found" hours of eight executives and cut one of them from the staff — a cost-displacement benefit. Either way, it represents a benefit. So even among executives, the Booz Allen study implies that OA appears to pay.

Recently a new type of executive measurement, performance accounting, has gained recognition. A leading exponent again is John Connell. He was asked to explain it during an interview near Office Technology Research Group headquarters in Pasadena. Specifically, what form does this new accounting take?

"You have to apply certain new principles," he replied. "In addition to the normal debit-and-credit financial principles you now have, you must 'account' for performance. This isn't so strange — it's used in manufacturing and marketing all the time. There they account for sales by volume, sales by territory, all of which are based on performance. These don't have a thing to do, really, with the revenue flowing in and out of those sales."

Such principles, Connell said, must get moved over into the office along with another concept: the capitalization of human resources. "People should be looked upon as a capital asset for purposes of performance accounting even though they can't be for purposes of financial accounting."

Asked for an example or two of where these principles were being used, the OA user-group director said he knew of no organization where the approach was fully developed. "But I know of a few firms beginning to struggle with it," he went on. "Merck and Upjohn have done extensive work to develop financial models for human resources accounting which calculate the capital impact of a particular investment involving people. And they come up with an ROI, a return-on-investment result that is treated by management the same as if they were investing in a machine for the plant."

Given the example of an investment in electronic mail, Connell painted the following scenario: "They would go through a series of calculations using their model and then say, 'We estimate this system will have the following impact on the people using it. Here's the way it will change the number of people we need, the skills they'll require, and the growth curve of our human resources asset.' They can also calculate how well that asset can perform in meeting the sales projections that the marketing people have come up with. At Merck, management will accept that with the same level of credibility as it accords the vice-president of manufacturing when he comes in with ROI calculations for a new machine."

The "Art" of Management

While work goes on at centers like UCLA to hone such techniques, and while human resources accounting has also been used to justify corporate takeovers in which the price paid for a company far exceeded "book value" (the value of its human resources making up the difference), these new techniques have not been welcomed with open arms in traditional accounting circles. Nor are they yet appreciated by most facts-and-figures managers. However, it might be said that some of them do not understand the modern office, either, as the strong competitive resource it has the potential of becoming.

All the same, in cost-justifying OA's softer benefits, one ought not pass lightly over other concepts so beloved of business traditionalists, like intuition and the "art" of management. Even such a realist as John Diebold, whose book *Automation* gave the word to the world as far back as 1952, believes there comes a point beyond which efforts to justify OA make little sense.

"I think you *can* measure it; it's very difficult," Diebold said of executive productivity in a 1982 interview with *Administrative Management* magazine. "You could spend a lifetime trying to quantify it, but I'm not sure that's a good use of resources."

In assessing OA, a manager "has to be able to *recognize* that he can get material changes in productivity," the noted New York-based consultant said (emphasis added). "If you can get a financial analyst to solve a problem in an afternoon versus a week and a half by use of these systems, you know you've got a big yield in productivity. If you can get a machine designer to come up with an optimum manufacturing combination for a product in a day or two versus a month and a half, you know you've got a big yield in productivity."

An OA planner can be pulled to and fro until dizzy by the argument and counterargument over how "hard" cost justifications must be. "It's absolutely irresponsible not to build a financial case for any installation," according to Vincent Pica of E.F. Hutton Co., quoted in a supplement, *Office Today*, of the *New York Times*. "Without justifying the cost, how can top management make an informed decision about moving ahead with office automation?"

"The tricky question of defining and justifying costs for OA equipment has troubled business for years," says Omni president Randy J. Goldfield. "While it is difficult to put price tags on many intangibles, like better, faster decision-making, it is necessary."

But Michael Hammer of the Laboratory of Computer Science at Massachusetts Institute of Technology rejects the notion that anything less than direct financial measurement is also less than useful. "'Hard' dollar savings are no more reliable a basis for justifying office automation than any others," he told the 1982 Office Automation Conference in San Francisco. "Nor are they more meaningful. The measures that the management of a unit feels are most indicative of its performance are usually the most appropriate ones to use."

What cost-benefit analysis ultimately must rest upon, according to Hammer, is "the essence of management" — namely, making decisions in the context of incomplete information. "The [manager] must examine the anticipated costs and returns, evaluate the accuracy of this information, and make a judgment call on whether the anticipated benefits are sufficiently likely to justify the expected costs," said Hammer.

No Big Problem

The buck may stop — and the bucks may start — with management, but the question of OA cost, at least unexpectedly higher costs, did not loom as a major problem for the AMS Foundation survey group. Only nine percent called *higher than anticipated costs* a major drawback during planning or implementation. Only two respondents mentioned cost at all in reflecting on their OA experiences. "I'd make higher cost estimates," said one. "Quicker results at less cost would have pleased our board more," observed the other.

Perhaps another confidence-building factor, beyond confidence in one's own management judgment, is the prevalence of success stories reported in OA literature and from conference rostrums. No two cases are alike, which explains why so few rules-of-thumb exist for gauging *X* result from *Y* investment. But here and there nuggets of statistical gold can be gathered, like Poppel's finding that executives spend 68 percent of their time communicating, of which 40 percent is in meetings and the rest on the phone, and planners would do well to collect them.

Other relevant statistics include:

- *Secretaries' time:* 40 percent typing, 60 percent administrative (useful, perhaps, for a WP assessment).
- *Average number of interruptions in a manager's day:* 15 (an argument for electronic mail or voice mail, so that not all calls need be taken the instant they arrive).
- *Percentage of business phone calls that do not get through the first time:* 50 (another electronic mail argument).
- *Average number of phone calls to schedule a meeting for 10 people:* 35 (check that calendar-management software).
- *Cost of a dictated business letter, including principal's and secretary's time:* $8.10 as of 1984 (definitely an argument for WP).

As for WP itself, many OA professionals say it is so well proven by now that management can simply accept the premise that it "works." To subject WP to costly investigations and tests would be a waste of resources.

Again it comes down to management judgment, because even here one encounters opposing arguments. But we leave them for another chapter (Nine) which deals with pilot projects, consultants, and other sources of supportive information.

Chapter Four

WHO PLANS?

With tongue perhaps slightly in cheek, the head of the Office Automation Society International (OASI) described the ideal leader of an office automation planning team like this:

"An innovative manager, a technologist, a cost accountant, an organization specialist, a catalyst, a doer, a planner, an educator, and an ergonomics specialist. In addition, he or she must like people. . . ."

As quoted in *Computer Decisions* (June 15, 1984), Paul Oyer, OASI president, summed up the requirements with one job title: "Jack of all trades."

Jacks and Jills of all trades are not so easily come by these days, especially when the trades embrace so many skills and professions. So it is not surprising that OA planning involves many different participants. What is surprising is the frequency with which organizations put the wrong leader in charge of the participants and wind up with no leadership at all.

The Right Leader

The need to look at the big picture and plan long-term runs insistently through the comments of OA planning veterans. However, the *ability* to look at the big picture is another matter. While it is hazardous to fit people into stereotyped roles, it is fair to say that management generalists are often the ones with the greatest understanding of overall organizational needs.

Yet, because OA so plainly involves technology, and because technology, despite its high cost, so often "doesn't work," someone with a record of managing systems is commonly put in charge of the OA project as insurance.

To say such a choice is wrong would be patently unfair to those DP and MIS executives who today work hard to address the needs of an office environment vastly different from the computer shops in which they began their careers. Yes, they understand technology, but they understand *their understanding* in the context of changing departmental relationships, new communities of end users to be served, and the business aspects of their organizations over the long run.

To say that putting OA planning responsibility into the hands of a generalist is right depends entirely on who the generalist is. Has he or she kept up professionally? Is that individual truly as holistic in outlook as OA needs demand? It may hurt to admit it, but the world of administration contains its share of "managers" who have never gotten much beyond cleaning-and-maintenance contracts.

Still, the following advice prevails:

- "Plan more from a strategic viewpoint" — vice president of information systems, Illinois insurance company.
- "Put together your strategic plan first. Know your organization and its goals" — associate administrator, Pennsylvania healthcare center.
- "Look at the big picture, then plan as far ahead as you can" — vice president-administration, Colorado insurance company.
- "Develop architecture before implementation" — manager, Alberta energy company.

From the above, it is clear that more than a question of "who's in charge" is at stake. There is an implication of avoiding haste, of not putting the cart before the horse. A theme throughout the comments of the many executives contacted for this study is "allow ample time." Too often, managers rush out to purchase computers and only later try to figure out how to fit them in. (For more on planning timetables, see Chapter Five.)

Generalists or Specialists

Who *is* in charge? Judging by titles, it would appear that management generalists outnumber systems specialists two to one as leaders of OA teams. Going again by titles, equal numbers of generalists and specialists serve as team participants. (Titles of all OA planners covered by this study appear in Appendix B.)

Not all planning is team based, however. One out of two organizations goes the one-person route. This conforms with an earlier Omni Group finding, wherein half of all small companies surveyed put OA development into the hands of an "office manager" (the title used) or the president. Among these firms of 100 employees or less, only 21 percent assigned OA planning to an interdepartmental committee; another six percent relied on single departments with "OA responsibility." Among firms of *Fortune* 500 caliber, however, these proportions reverse. Fifty-five percent assigned planning to OA departments or interdepartmental committees; only 14 percent went the one-person president or manager route.

The value of the "office czar" as opposed to a planning committee can be argued endlessly. As both the Omni and AMS Foundation studies indicate, the choice may be influenced more by company size and style than by anything propounded outside.

Even so, few in the OA world speak out for going it alone:

- "Effective OA systems require a team approach" — Jean Greene Dorsey, deputy director of MIS, City of New York.
- "No one person or function should plan for the automated office" — Gad Selig, former OA planner with Continental Group, now an OA consultant.
- "The need [in OA] to better understand people and organizations on a holistic basis requires a multi-disciplinary approach" — Eugene Manno, Office Management Systems Division, Honeywell Inc.

Yet, planners who have done one-person planning cite its practicalities. It is more direct; there is less compromise with differing points of view; and besides, what's to prevent the "czar" from calling in advisors from time to time?

Looking to 1990 and beyond, one can see two lines of development and two schools still at loggerheads over this question. On the one hand, OA will have become so advanced and tied into so many more aspects of business than now, that interdisciplinary planning would be the only viable course. On the other hand, OA will have become so advanced that it would be commonplace in business. What need will there be to intensely discuss what has become proven methodology that needs only slight adaptation to fit a particular workplace?

Until the question is resolved, the committee approach is, for most organizations, the safer way to go.

No One in Charge?

Perhaps most revealing in the AMS Foundation survey is not the assignment of leadership responsibility one way or the other, but the apparent, sizable lack of leadership at all.

In half of the cases involving committees, respondents failed to indicate which of the members served as chair. That this was no oversight can be seen in answers to question like "What advice would you give others?" and "What would you do differently?" — questions asked in several ways throughout the survey. Typical answers included:

- "I would establish a clearer line of authority and responsibility" — manager of administrative services, New England-based restaurant chain.
- "I'd consolidate into one office the staff which does the [planning]" — program manager, Air Force accounting and finance center.
- "Clearly establish who's in charge" — administrative vice president, midwestern law firm.

From other comments and case histories, it is evident that service on planning committees, and even the calling of meetings and tasks, too often has a nebulous quality. Frequently this on-again, off-again drift coincides in hindsight with comments about the lack of adequate top-management support. Where senior management does maintain strong interest, things get accomplished (see Chapter Six).

Outside Consultants

Thirty-five percent of organizations in the AMS Foundation survey called in outside consultants to help with OA planning. Many who didn't wish they had:

- "We needed more technical expertise in DP and communications on the project team."
- "Doing it again, I would obtain more information on software and also more knowledge of networking."
- "My advice? Use independent, reputable, outside, paid consultants."

This perceived need for outside expertise is mirrored throughout the survey in comments that pointed to inadequate resources. There was "not enough" funding, time, in-house expertise, vendor support, or management understanding. Even so, for every kind of gripe, someone registered a countergripe. Said a Virginia school administrator: "We should have spent less on the consultant and concentrated more on our system." And a manager in New York government snapped: "Never trust vendors."

Leadership Qualities

Regardless of career background or special expertise, the trait that most befits OA planning leadership is just that — leadership.

Every OA effort aims at reaching management goals through systems involving technology and people. If a project leader lacks understanding of any of these elements, he or she can presumably obtain it from others. What cannot be begged or borrowed is that inner "can-do" confidence which presses top management to more clearly spell out the goals to be won, if at any time they are not clear; to lean on vendors to come up with better solutions than they have; to motivate team participants to dig deeper for technological answers, if that is the need; or to develop more effective training fare, if that is what's missing.

All of these characteristics become more significant when the process shifts from planning to implementation, because the person who led the planning typically takes charge of implementation. The traits of persuasion and strength now play on a larger stage. The man or woman in charge must be a "people person" outwardly to end users and top management, and inwardly to a team with its own fears and interests.

As we have seen and shall see again, OA changes organizational relationships because it changes ways of working. Departments win, departments lose. Jobs come, jobs go. While it is important to involve as many interests as early in the planning as possible, OA executives should never forget the inner, conflicting feelings of the various people seated so placidly around the table.

In one of those "what-they-say" versus "what-they-mean" jokes, the DP manager says to an OA planning session: "To provide end users with the benefits of distributed processing, we needn't abandon the advantages of centralized organization." What he means is, "I've got a good job in that center and damned if I'm going to let you jerks louse it up!"

It takes sensitivity to these hidden motives, plus persuasiveness and forcefulness (whichever does the job), plus an ability to express the ends and means of the project, to channel the considerable energies and knowledge of the group into purposeful action. This is leadership.

"The leader's most important role is that of facilitator," says John Connell. "The mission is to educate."

The previously quoted Air Force program manager may have summed it all up best. "OA is a complex undertaking and requires a balance between technical and human factors if the automation is to really work in a day-to-day environment."

When OA "really works," people perform better and business produces more effectively. But no one ever said that *making* it work was easy.

Chapter Five
HOW MUCH TIME IS ENOUGH?

A planning time of one year is the midpoint in a sample of recent office automation projects. Forty-five percent of the organizations surveyed for the AMS Foundation survey took 12 months or less to get from the start of OA planning to the start of implementation. For another 45 percent, planning took a year or more (see Table 5-1). The remaining 10 percent said they were currently planning but had not yet begun to implement.

Many planners, reviewing their projects, said they wished they had more time. A Denver-based administrator whose planning cycle lasted 10 months was typical. "Allow more time," he said. "Market conditions are very unpredictable and new products emerge all the time. To be tied to a deadline may result in choosing equipment soon outdated."

"Study more up front," advised an administrator at a Louisiana chemical firm where planning lasted more than a year.

On the other hand, a methods manager with a West Coast maker of pharmaceutical instruments where planning spanned 26 months would have "speeded the decision process."

Add-Ons Go Faster

There are projects and there are *projects*, and it may be assumed that efforts to plan brand-new systems take longer than those which expand an established base. According to survey data, this assumption is correct. Half of the projects in which planning took less than a year involved additions to existing systems. All of the projects taking 18 months or more were start-ups.

Not all situations afford the luxury of time. Haste may be one of the goals top management sets if business reasons exist. Such was the case of the Revenue Cabinet, or tax department, of Kentucky in 1982. It was burdened with archaic systems that were causing the state to lose an estimated $100 million in uncollected taxes each year. Constant overtime work by cabinet personnel only deepened the problem.

John Y. Brown, Jr., then governor, issued a not-impossible order: "Automate." He then added a near-impossible condition: "Do it in three months." It took four. But what cabinet secretary Roland Geary, OA section supervisor Scott Bartelt, and others accomplished in that time gives vivid support to the notion that OA crash programs can succeed when sufficiently focused, and when top management stands behind them.

In a swift, intensive study, the cabinet team contacted everyone from secretaries to executives to find out where OA reforms could best

Table 5-1
From Planning to Implementation — How Long?

Elapsed Time Between Onset of Planning and Onset of Implementation	Percent of Respondents
Less Than A Year:	
Six months or less	20
More than six months to a year	25
	45
More Than A Year:	
More than a year to 18 months	20
18 + months to two years or more	25
	45
Planning But Not Yet Implementing	*10*

be applied. Then, with needs established, they set out to find equipment that could best fulfill the needs. Here, two criteria were obvious. Because of the relentless deadline, the equipment had to be fairly easy for users to learn. Also, the vendor had to be well set up in Frankfort, the state capital, so support could be assured. Wang met the requirements better than any competitor, and got the contract.

In little more than 100 days after the governor had given his order, five different, yet compatible, systems arrived at five cabinet buildings in Frankfort. Four systems were linked via phone lines; the fifth stood alone. Other links tied these four to remote sites for dictation and to the state's IBM mainframe, which in turn accessed property data in county offices. With all that power and reach, the new $500,000 integrated system was not only able to speed the collection of revenue, but also handled legal searches, statistics, and word processing.

As *Office Administration and Automation* reported two years after the launch, worker productivity had increased by more than 75 percent, overtime was practically eliminated, and a costly backlog of paperwork had been reduced to almost zero. The only people not pleased were Kentucky's erstwhile tax evaders, who now more readily pay what they owe.

The Long View

In contrast to cases like Kentucky's, a number of firms plan strategically for advanced office technology five years or more into the future, and tactically for new systems a year or two down the road.

Atlantic Richfield, while not yet implementing anything major, keeps its "OA vision" attuned to a five-year-distant horizon. The exercise gives management "an idea of where we're heading," says planner Allen Smith, "even if we can't reach our destination right away. The more we learn, the quicker we can take advantage of solutions when they do arrive."

Many OA executives, including several contacted for this work, echo that theme. "Start planning now," says the quality assurance manager with a voice-mail company, who implemented the firm's OA system. "This is the problem-prevention phase" — even if management hasn't given a go-ahead.

Other OA watchers caution, however, that too much emphasis on the long term, with no active follow through, can become an excuse for procrastination. "We're waiting for that next technological break-through" really means, "Don't bother me with OA — I've got enough problems already."

Robert Kalthoff, president of Access Corp., a Cincinnati-based sys-tems integration firm, points out that planning costs money like any-thing else, and while phrases like "managing the change process care-fully and thoroughly" sound good — and are good — one can extend a good thing too far.

Kalthoff estimates it costs between $15,000 and $20,000 per month just to plan for and negotiate an office system in the quarter-million-dollar range. Obviously, at those prices, planning ought not last longer than it has to.

As a result, OA advisors often suggest pursuing two timelines simultaneously — a one-year, short-term line, and a multi-year line. "Keep your head in the clouds, but your feet on the ground," says consultant N. Dean Meyer, founder of the Society of Office Automa-tion Professionals (SOAP). Meyer also disputes the notion that plan-ners should hold back until they see what the next generation of tech-nology will bring. Planners already know that new systems will cost less and have greater processing power. And while the products may be packaged differently and some will perform differently, many will still address office tasks that, in themselves, cannot be expected to go away.

Long- and short-term planning should not be seen as conflicting, but as complementary, says former Exxon office-systems manager Robert Dickinson. "A broad view of where you're going to be in five years is absolutely essential to the development of a one-year plan," he says. "A one-year plan is no good without the perspective [of] a long-term outlook."

SRI International consultant Alexia Martin speaks of "planks" in a long-term platform which, over time, can be tested, reshaped, or torn out. The planks are general statements of long-range corporate goals and likely office systems, written with as much foresight as can be mustered at the time. The statements may not survive later knowl-edge, but when posted for all to see, they serve as bases for considered decisions. "Short-term implementations," Martin says, "only enable an organization to combat immediate problems; implementations carried out under the umbrella of a long-term perspective will ensure continued business success."

Among the AMS Foundation survey group, the view is fairly long term. Although most respondents do not project OA growth plans beyond three years, 42 percent look four years or more into the future. Four percent scan a horizon at least seven years distant. No one admitted to addressing only immediate needs.

Managing the Time-Line

While the general view of OA planning puts corporate needs and "givens" on a calendar of progress in which these situational issues gradually take the form of "solutions" (new systems and procedures), specific formulas exist for managing this process phase by phase.

A three-phase model was presented in *Today's Office* (November, 1984) by Bill Beers, a manager and instructor at the AT&T Information Systems Institute for Communications and Information Management. While the phases — strategic management, needs assessment, and design — may seem obvious by now, Beers divided each into a set of discrete steps to help planners find their way through what is a holistic undertaking as much as a linear process.

Strategic management begins with a determination of business goals. Anything the planners consider later with regard to technology must support those goals. This, too, sounds obvious, but Beers argues that planners often ignore this relationship, repeatedly isolating dramatic "OA benefits" from quickly forgotten original objectives. (Goal-setting strategies are further discussed in Chapter Seven.)

After goal-setting, the proper next step is to plan and prepare for a project team, then select its members and give them their marching orders.

Phase two, needs assessment, basically answers questions like: "Where are we?" and "What kind of system will best achieve our business goals?" Here, says Beers, planners should examine applications that show the most potential for a decent return on investment. This is a basis for cost justification. "Contrary to what some believe," he says, "cost justification is still a key implementation criterion, and it must be done."

Part of planners' work at this stage is to come up with specific methodologies for answering "needs" questions. Beers names questionnaires and computer-assisted cost and office-work analyses as useful approaches here. He says that once teams get a fix on organizational needs, "they confront the question that really controls decisions on OA: Where are the organization and technology going?" Says the

AT&T executive, "There are no quick and easy answers here."

The best answer, Beers says, echoing the views of Meyer, Martin and others, must be to build systems around the long-term goals of phase one. Again, many managers fail in this, choosing technology without regard to future, and sometimes even present, company needs.

In the design phase, planners produce both a systems blueprint and a game plan for introducing it to users. "The key is to develop an integrated design strategy, not just to purchase a disjointed collection of office equipment," Beers states. "An integrated strategy is characterized by compatibility, open architecture, and provisions for growth."

Elements of such strategy include training, ergonomic studies, and assessments of whether or not goals are furthered and people perform better — no small piece of work. To cut this down to manageable size, planners might take an additional step, to the pilot test and studies of how technology-supported executives use time. However, Beers cautions that, realistically, the only way executives can be measured is by the quality of their decisions.

A Ten-Year Strategy

At one of Office Technology Research Group's semi-annual meetings, OA planner Gad Selig outlined a strategy for introducing advanced systems that had four phases, and looked 10 years into the future. That the plan was never fully adopted by the organization for which it was prepared is less important now than the long-range approach it exemplifies.

Some of this may seem obvious and repetitive of other plans, but this may also be an indication that an accepted OA planning method is quietly evolving, and that the people who say OA will be a relatively straightforward, "proven" concept beyond 1990 could be right.

Phase one in Selig's outline, lasting about a year, involves the get-ready activities of "initiation and exploration." Senior management states the corporate goals which will guide planners toward particular systems solutions.

Phase two, "migration and expansion," sees systems implementation underway in what is frankly expected to be a rough three years. End users become technically dependent participants of sorts, but not really comfortable with the equipment. "Uncoordinated proliferation" prevails among various functions like word processing, electronic mail, and electronic filing. Unclear jurisdiction and power struggles hamper management control.

In phase three, "consolidation and formalization," (years five and six), things get better. A more knowledgeable planning- and control-oriented style of management replaces what Selig called the "lax" style of the first two phases. Upgraded systems involving executive workstations, digital networks, and voice- and touch-entry methods coincide with the start of integration among applications. (The voice-entry hope was clearly premature.) End users now eagerly participate, focusing more on what can be done and less on how to do it.

Phase four, "maturity," sees users fully backing OA and management understanding that profits derive from information as a business resource. Increasingly, investments are redirected into "intelligent" assets. In these four years, planning begins on a new decade of OA development.

A New Order

The importance of these scenarios is not their step-by-step progression, but their representation of a new order of responsibility in the management of offices.

Nothing like this ever existed before in the field of administrative management, even though the need to cope with change was never absent. New systems would replace old systems; departments would regroup. Even office lifestyles had to adapt to the social turmoil of the 1960s and the women's rights movements of the 1970s and 1980s.

All of those were fairly definable, addressable challenges, however. They were episodes, not journeys to the unknown. Planning for OA, at least ambitious OA at the present state of its art, puts the challenge of coping with change on a newer, higher level — indeed, on several levels simultaneously.

What OA planners undertake is nothing less than the introduction of revolutionary technology into a workplace that has known only evolution. It takes time to prepare for that; time to slow the impacts of revolution to a pace of human acceptance; time to weigh likely benefits from untried and expensive systems on the soft scales of performance and other intangibles, such as the value of the information these systems so swiftly deliver.

More than time, management acumen, human understanding,

and leadership are needed. And where upper management appears remote, misinformed, and even hostile, large reserves of inner strength and courage are required.

Chapter Six
THE ROLE OF TOP MANAGEMENT

"They talk about it, but. . . ."

The administrative services manager at a central Canadian agricultural agency paused meaningfully in response to a question on the role top management had played in fostering office automation.

"Meanwhile," he added, the mood shifting, "I do as I feel I can, and as I see the need."

A good situation? Not in the eyes of many planners, whose regrets about their OA experiences center on not having senior management sufficiently involved, and on not having the ground rules of the project sufficiently spelled out. The following planners' comments note what they would do differently.

- "I would enlist greater management support up front."
- "Next time, I'd get a commitment in writing as to what was actually required."
- "I'd get top management commitment and [agreement on] methodology and justification rules in advance."
- "I'd gain high-level sponsorship for the OA study, and use that sponsorship to get user participation."

Their advice to others is much the same:

- "Obtain senior management commitment."
- "Get top management's commitment to financial support and standardization."
- "Obtain a complete go-ahead from top management so they can't sit back and say, 'I told you so.'"

Should Top Management "Plan"?

Top management support is one thing; active involvement on the planning team is something else. Consider the following "What would you do differently?" comments:

- "I'd not allow top management to override user preferences as to hardware and software suppliers."
- "I'd insist on less outside interference, especially top management's."

Tales of careful user analyses and vendor-preference studies being overturned by "country club agreements" between the boss and his or her golfing buddies, who just happened to work for a different vendor, run sourly through the annals of OA defeats.

Tales of basic misunderstandings of OA's role — where management support often amounts to a notion that "if others are automating, we'd better get computers, too" — rumble darkly over drinks at the close of OA conferences. A few examples:

- "OA capabilities still not fully recognized."
- "Top management set unrealistic goals, has unrealistic expectations."
- "Top management alternately complains about activities being computerized for no purpose (i.e., for the sake of computerization) and then expects us to do everything without regard to compatibility and integration problems."

In small organizations, where the president must wear many hats including that of office administrator, he or she almost certainly will engage in OA planning and implementation (see Chapter Four). However, in organizations of any size, most OA observers agree, the boss should be committed to the project and visibly supportive, but except for rare appearances, ought not take part in planning.

Support Runs High

Happily for most OA executives surveyed, the degree of involvement by top management appears to be about right. The executives in the AMS Foundation study were asked to check one of four statements which most closely fit their own OA planning situations. Thirty-four percent chose "Top management was personally and actively involved throughout"; 45 percent opted for "Active and involved at the beginning, and continues to show interest through periodic briefings" (see Table 6-1). This assessment shows that the interest of upper executives remained strong in almost four-fifths of the cases.

Significantly, initiatives to pursue OA do not come from the top every time, though they usually do. The study, seeking only titles, found the prime movers at times to include a vice president, controller, administrative manager, information systems vice president, information systems director, and an associate director of data processing.

Relationships

Where planning went smoothly, OA executives used various means to keep upper echelons aware of developments. "We brought top management to a computer comfort level by introducing them to personal computers in anticipation of the system," says the administrator of a nationwide service organization based in New York.

"Executive 'sponsors' were presented with progress reports and recommendations but were not active participants," reports Karl Hellwig, C.A.M., administrative services manager at Foremost Insurance Co., Grand Rapids, Michigan.

Table 6-1
Top Management is Involved

Top Management Was:	Percent of Respondents
Personally and actively involved throughout	34
Active and involved at the beginning, and continues to show interest through periodic briefings	45
Involved at the beginning, but interest has waned	14
Never involved	7

Note: Respondents were asked to check the one description that most closely fits their situation.

In other instances, contacts and guidance were "top-down." A notable case involved offices of the New York State Senate in Albany, where, in a political environment, "the intense personal involvement and support by the Secretary of the Senate was largely responsible for success." The source wished not to be identified.

Sometimes the degree of involvement changed abruptly. "Our ini-

Planning and Top Management — Some Views

In addition to assessing top management's role in OA planning for Table 6-1, OA executives were asked for comments on this general subject. Here is a sampling of their replies. Other comments appear in the text.

- Initiative came from associate director of data processing. Top management awareness and support has been slow to develop. They are supportive, but in a passive way. The driving force has come from below top management.

- Top management consists of general manager and four assistant general managers. Initiative came from the assistant general manager for administrative services and the DP manager. The general manager endorsed the concept and keeps current by way of periodic briefings on direction, capital expenditures, and the like.

- Top management was and continues to be very supportive of OA efforts.

- Top management authorized a study group and financial commitment after a determination of benefits as presented by the general manager's level. Follow-up has been left to the general manager and individual managers.

- Presently a three-year commitment has been made by top management. A task force is led by the administrative vice-president.

- Initiative came from the VP for information systems.

- Top management has been active only during early evaluation of WP equipment. OA *per se* has only received casual attention.

tial plan was the result of a very involved CEO," a school administrator reported. "However, a change in this position has resulted in somewhat lessened interest." This is perhaps a delicate way of putting it, but the influence and importance of management support are plainly conveyed.

Top management, of course, plays more of a role than "support." Top management has the often-demanding initial role to enunciate clearly the goals and purposes toward which OA can help the company advance.

In short, top management directs. The form that direction takes and the expanse of its charter are matters we turn to next.

Chapter Seven
SEEING THE "BIG PICTURE"

Think of OA strategy as a chemical process. At least three management elements interact in the crucible of planning. How much of each you add to the mix helps determine the compound result.

The three conceptual ingredients are:

1. *The goals top management sets.* Goals were discussed in Chapter Two where a distinction was made between value-adding and cost-reducing objectives. Various systems architectures — end-user and centralized — designed to gain those objectives will be covered in Chapter Eight.

2. *The "top-down" or "bottom-up" strategy one uses.* Are systems and their implementation largely decreed by management? Or can departments, and even individuals, pretty much "do their thing"?

3. *The scope of the environment in which planning takes place.* In other words, how big is the picture in which OA will occur? Is the scale departmental, organizational, or does one "see" an entire field or industry?

In the office world at large, despite businesslike demeanor and talk of productivity, most OA systems "grow like Topsy." Karen Orten, who has studied OA growth in over 300 large companies, and has identified four basic OA strategies, says the *laissez-faire* model — the *absence* of strategy — is probably most common.

Laissez-Faire

Orton, vice-present for microcomputer education at National Training Systems, Santa Barbara, says individuals in a *laissez-faire* environment simply use discretionary funds from department or division budgets to buy whatever hardware or software seems useful to them.

Something less than "better business" or "greater productivity" on an organized scale thus drives these moves. Something rather like this computer here, this printer there, this cable connection somewhere else, seem like good ideas now. It is not surprising that such unrelated "good ideas," bottoming up, so often occur in organizations where no one sees any context beyond what is right before their eyes.

If there is anything to be said for *laissez-faire*, it's that it addresses needs quickly — the fast fix. A problem down the road, of course, is the lack of compatibility among such a potpourri of products, which is a major impediment should a company wish to network purposefully. Meanwhile, management has only the vaguest idea of what is being acquired, because it doesn't watch. Orton tells of a major company which guessed it had a dozen or so personal computers on site. A count revealed 312 in actual use.

Mid-Course and Top-Down

Between the almost total non-strategy of *laissez-faire* and the organized, mandated approach of top-down, Karen Orton found two other common policies which blend elements of both extremes.

One she calls the "approved-vendors-and-sources" approach. This is *laissez-faire* with the stipulation that employees stick to one or two agreed-upon brands of merchandise. Companies like Hughes Aircraft and DuPont allow workers to "automate at will" this way. Some firms maintain in-house computer stores. In Washington, the General Services Administration runs one for the Federal government.

While this strategy allows eager beavers to automate at their own pace and controls their actions a bit, its drawbacks are similar to that of *laissez-faire*: a payout limited to small personal jobs and the possibility down the road of nonstandard skills, files, and equipment.

More controlled is "task-driven" strategy. Orton calls it the most prevalent OA approach next to *laissez-faire*. It addresses operational needs, usually on a sectional or other close basis. It is relatively easy to justify because the task involved is usually discrete enough for planners to calculate some benefit. It may even be implemented as part of some larger OA scheme. Where it is not, of course, those same incompatibility problems could later obstruct efforts at departmental integration.

The top-down model is top-of-the-line. It has the elegance of totality, or at least comprehensiveness. The gears of corporate strategy mesh well with systems strategy and integrated technology throughout the organization. Here you know who's in the driver's seat — top management. But the vehicle is slow. Middle managers and other office professionals grow anxious and impatient at the often long delays between the commencement of planning, the announcement of decisions, and the delivery of equipment. "In the long run, though," says Orton, "this strategy is likely to get the best results of them all."

The Wider View

In contrast to OA's widespread, Topsy-like growth, the projects examined in AMS Foundation research for this book have a strong top-down quality about them, as indicated by the high degree of top-management involvement (Chapter Six). The fact that for half the group planning lasted more than a year is another indicator of top-down direction (Chapter Five). And the broad environmental context in which planning took place is convincingly yet another, as shown in Table 7-1.

Here again, OA executives were asked to check the one statement out of four which best described their situation. "Before dealing with the specifics of OA systems," they were asked, "what larger issues did the planner(s) consider?"

Significantly, three-fourths indicated that planners strategized beyond the confines of any one department to at least an organizational framework, if not to their entire field or industry.

What does that mean — to strategize across an industry? As one writer has described it, it means positioning one's organization within its business universe and estimating, as best one can, where both the organization and that universe are going.

At this stage, it has nothing to do with offices. It has to do with articulation of purpose, of where the company (or agency) wishes to be, the pace it wishes to set, the objectives it wants to accomplish.

Only top management can truly state these "grand know-whys," as

Table 7-1
How Big the 'Big Picture'?

Question: Before getting to system specifics, what larger issues were considered?	Percent of Respondents
We strategized long-range about our industry, or field, our competitive environment, our organization and its needs	42
We looked at our entire organization, but did not consider outside factors like competitors or industry	35
We looked at one division or section, but tried to foresee its future needs in a broad context	20
We recognized we had specific problems and went for quick remedies	3

Note: Respondents were asked to check the one statement that best described their situation.

they have been called, in contrast to the "know-hows" which may achieve them through OA. Such factors include:

• who we are;
• what the economic climate looks like;
• what our market looks like; and
• what our competitors are up to.

With this information, the organization then says, "Here's an area of uncertainty. Here are strengths we can maximize. Here's the kind of organization, if we work at it, that we can be in five or 10 years' time."

While this represents a very big and very general picture, at least OA planners can see a purpose, get some bearings, and can more confidently get on with their mission because they perceive where *forward* is.

Now they can address follow-up questions, still on big-scale terms. What kinds of informational needs and service needs do our staffs and managers have? What kinds of communications do they engage in? And — back to strategies — do we want to free up their time so they can perform new and better work, or are we really looking to hold down costs, or cut them? Only now does the office and the designs of systems it can have come into focus. Now the technology that can fit the designs that help advance the strategies can be appraised and tested.

In fact, 42 percent of the surveyed executives identified with the statement, "We strategized long-range about our industry or field, our competitive environment, and our organization and its needs."

"Put together your strategic plan first — the purpose of the organization and its goals," said one such respondent, a Philadelphia-area health-care center manager. "Then ask how OA can meet those goals in all its forms."

From Plan to Project

Getting from strategic plan to purposeful project poses the following basic question: After all the wider views have been seen, how do planners focus on what OA can do to meet the goals? In other words, what do OA planners do — specifically?

Many recipe-like checklists appear from time to time, and while they can be useful, they often meet with "my-case-is-different" objections. The fact is, OA needs and OA solutions are as varied as the purposes and procedures of business itself. In a strategic overview like this, it's not possible to answer — specifically.

Gilbert Konkel and Phyllis J. Peck, two Milwaukee-based OA authorities (one a manager of office services for Arthur Young & Co., the other an in-house consultant for Allis-Chalmers Corp.) have published a reasonably general outline of how the turning point from strategic long view to short-range problem solving can be negotiated successfully.

Their seven "problem-solving steps" involve:

1. defining the basic problems from which present "symptoms" radiate;
2. setting down the criteria against which proposed solutions will be judged;
3. brainstorming the proposals;
4. evaluating them as to costs of money and time, ease of execution, potential consequences, and probable effectiveness in meeting the criteria of Point 2;
5. choosing the best solution;

6. implementing it; and

7. evaluating it for any necessary later adjustments.

Fine-tuning even more to manage the time line on which these steps occur, Konkel and Peck divide the project into three sections. The first part includes meetings with management, orientation meetings with the staffs of sections to be studied, gathering and analyzing information about their work (done possibly with the help of vendors and consultants — see Chapter Nine), and the preparation of preliminary equipment proposals, space plans, and reports to management. The second phase involves budgets, the awarding of construction and furniture contracts, and the acquisition of equipment. The third phase involves training, leading to start-up day when the system (or at least the first parts of it) are put to productive use.

Chapter Eight

SYSTEMS ARCHITECTURE

At Home Box Office, giant of the pay-TV industry whose programs reach more than 12 million homes in 50 states, development of office automation has been almost as swift as the growth of HBO itself.

As recently as 1982, "OA" consisted of one Wang OIS 125, a small system with four workstations and two printers used mainly for word processing. A short two years later, 29 user groups were being served with close to $600,000 worth of WP-oriented hardware and software, plus rented and leased equipment. By 1985, plans were underway to link more powerful Wang OIS 140-3's in nine regional offices to the company's IBM 3083 mainframe at New York headquarters. Here, links already existed between the 3083 and a Wang VS100 minicomputer, the latter also connected to a tenth 140-3 by a Wangnet local area network.

Described like this, the HBO system may seem a bewildering assortment of model names and numbers. Actually, a basic strategy has guided its growth: an OA architecture built on a WP foundation, but expanding now to also become an electronic mail and information system tied in with the New York mainframe. And it is essentially a two-vendor strategy — Wang for the "office" applications and IBM for mainframe "data" operations, although a Syntrex system in the legal department was recently tied to the LAN.

A Different Growth Pattern

A much different strategy guides OA at the Bayvet division of
Miles Laboratories in Shawnee, Kansas — growth based not on WP
but on direct expansion from the mainframe. The system has been
growing steadily since 1977 when an IBM 4331 was brought in to cure
a sluggish order processing system at this major animal health and
veterinary products firm.

Order entry still lies at the heart of the system, but the vast network
which fans out from the mainframe today supports WP, electronic
messaging, and spreadsheet and research analyses at company sites in
19 cities of the United States, Canada, and West Germany. Thirty-
two Datapoint terminals and 36 major peripherals are interconnected
to the 4331 and to each other by various means — dedicated lines,
interface hook-ups, and dial-up and automatic dial-up WATS. In
addition, 13 of the units at an administrative center in Merriam, Kansas, work off a local-area Datapoint "ARC" network.

The HBO and Bayvet systems are exemplary not only for their contrasting patterns of growth but for the benefits OA has brought. Office systems group manager Karen Pardo at HBO and MIS manager
W.E. Foulkes at Bayvet both produce long lists of functions their systems handle: payroll records, government reports, market overviews,
contracts, engineering studies, scripts — a plethora of word-processed
documents required by their separate industries. Both cite savings of
money and time: more than $214,000 was documented at HBO's
Atlanta office in a six-month productivity study; another $381,000
was saved in one New York department, according to the same study.
At Bayvet, back orders were cut from 50 percent of volume to less
than five, delivery times were cut in half, and more effective collabor-
ation was achieved among widely separated executives, thanks to a recently added electronic mail facility.

Where to Begin?

The contrasting development of these systems makes yet another
point: that however different the starting place, one can "migrate toward OA" (to use a voguish term) eventually.

The mature systems of HBO and Bayvet are really not that far
apart. Both integrate WP and localized informational work like
spreadsheet analyses with the power and databases of a mainframe.

Both use (or will use) links of the underlying network for electronic mail and messaging. Yet both started from different functional bases.

Is any one OA foundation architecturally better than the rest? For particular firms, probably yes. However, as a general proposition, don't look for neat, pat answers, although there are some general principles.

Organizations with heavy document workloads — law firms, insurance companies, health-care facilities — might logically consider building OA on a WP or automated records-management base. Both are well-established functions; both have well-worked-out methodologies; and early payouts can be fairly well assured. However, both involve relatively high start-up costs.

Organizations (or departments) with heavy "knowledge" workloads — advertising, engineering, consultancies, marketing — might well go directly to end-user support, equipping professionals with their own PCs for bottom-up experience toward later, more formalized architectures. Consultants like N. Dean Meyer favor this approach, saying the important point is to get things moving. "Focus on integrating systems only after momentum has been built."

Furthermore, organizations with long-established DP operations might well migrate from these centers to management information systems for executives and then onto a wider array of end-user applications. Such firms have the resources to do it. Indeed, where DP and MIS are especially strong, where turf battles could erupt if any but these departments were put in charge, companies have gone this route if only to preserve political peace. Whatever the case, MIS systems, by cutting through masses of data and providing managers with understandable information, can generate one of OA's quickest payouts by enabling users to make faster, better decisions.

Build or Rebuild?

As we saw in Chapter One, the large midwestern utility with its AOSSG and AOSCT committees expanded both from WP and DP to a point where the new OA mission was to put them together. Monsanto, described in the same chapter, was also melding its WP and DP experience under MIS guidance to put OA tools directly into end-user hands. Rubbermaid, on the other hand, rested its "OA pyramid" firmly on a WP base.

A look at the OA migrations of organizations contacted for this study only reinforces the finding that no two situations are alike. If there is any pattern to the second and third phases of their OA development, it is a desire to link localized equipment to mainframes. Still, the cases are so individual as to preclude any common description.

An Iowa university seeks to integrate a Harris minicomputer with an IBM mainframe. A midwestern pharmaceutical company's focus is on the desktop tools "non-computer types could use to do their jobs better." A state insurance agency, having linked DP and WP functions, now has micrographics in the planning and research stage. A large temporary help firm seeks better management of its sizable database.

But as many organizations in the study are starting new systems as are building on ones that exist. A few of these are true start-ups — a first try at advanced technology. But more are re-starts — the replacement of older equipment (mainly WP) with stronger OA foundations. And some, like a Nebraska aeronautics agency, are departmental start-ups alongside existing systems: "We determined that with PCs we could handle our own work much cheaper than at central DP."

"Requirements of users and full production gains could not be met [with existing systems]," said an OA executive with the military. "In the past, automation was done on an individual, application-by-application basis, and so a variety of word processors, micros, and minis were implemented." A new OA system was in the planning stage as he said this, but implementation was not expected for a year.

"Primarily, we chose to begin with a separate standalone Wang minicomputer and also encourage the use of PCs," said another start-up planner with a midwestern insurance company. "As the technology matures, we'll interface these systems with our DP mainframe to facilitate communications and the downloading of information on corporate databases."

"Old WP equipment was phased out," an office administrator at headquarters of a retail chain explained. "The new OA [system] will interconnect with an existing DP mainframe but otherwise be independent." (See box for other comments.)

Three "Levels," One Office

Yes, each case is different, yet if one listens carefully, one can per-

Expanding the Present System?
Why? Why Not?

As many respondents in the AMS study said they were starting up new OA systems as said they were expanding systems already in place. Here is a sampling of the reasons they gave for taking their course of action. Other comments appear in the text.

Building on Existing Systems

- Central WP is in place — distributed WP as well. We'll use PCs to allow for distributed DP in the future.
- We'll integrate — or replace — Burroughs Redactron WP equipment with Basic Four "Business Office" and DEC equipment, which service researchers and physicians.
- We want to use more of our database.
- Goal of integrating with Harris 800 and IBM mainframe was instrumental in the planning.
- We have added components of telecommunications to existing DP and WP; micrographics is now in the planning and research stage.
- DP, WP, and telecommunications have been tacitly acknowledged as relevant. OA links to a mainframe environment and DP were planned [but] plan has undergone revision.

Starting New System

- Our present system is outdated.
- We decided to abandon our standalone Wangs and put in a more sophisticated system which would be compatible with present computer system. We obtained the IBM 5520.
- This is our first experience with OA. Plans are being made to extend it to underwriting and engineering department by mid-1985.
- Initial plans did involve our standalone system, but current pilot program involves mainframe-based WP. Territorial problems plus state-of-the-art condition of the mainframe led us to make this change.
- Existing WP systems offered little flexibility for growth and new applications.
- Our present system can't handle what we need.

haps hear a theme for OA designs of the future — even two themes, contrary yet coexistent. "Interconnect with a mainframe. . .otherwise be independent."

Dr. Raymond R. Panko, associate professor at the University of Hawaii's College of Business Administration, carries this dependent/independent idea even further. Office work, he argues, exists on three levels — individual, departmental, and organizational — and each influences, and is influenced by, emerging office automation technology.

At the individual level, work is extremely varied. In the past, to have supported each job with advanced technology would have been very expensive. But such tools are becoming increasingly "generic," Panko observes, capable of being tailored for specific applications more economically.

At the department level, where most OA efforts have focused, "it seems to make sense to distinguish between two basically different types of offices — Types I and II," Panko calls them. Type I's handle routine information processing chores, such as WP and DP, in highly structured environments. Type II's handle nonroutine knowledge work in ill-structured, complex environments. "The Type II office is the wave of the future," Panko states, "and we're just beginning to understand how to serve it. One thing is clear, however: needs in Type II offices cut across traditional boundaries" of the information-processing establishment.

The organizational level has not really been served by information technology, Panko observes, because "the tools to build integrated systems simply did not exist." That is changing, of course. Organizations now plan for and begin to implement integrated DP systems, OA systems, and databases. At this level, the biggest contributions of OA may yet be in the areas of voice and text communication, Panko says.

But the main point in discussing levels at all, he goes on, is not to continue treating them separately but to bind them in the "seamless environment" of future office supersystems. "Perhaps we should view the organization as being like a biological organism," he says. "Each office or department is like an organ — existing not for its own ends but to provide outputs to the rest of the firm. These organs work together as a result of processes that cut across them — such as nerve pulses or biochemical signals."

A New "Technostructure"

Projecting the analogy into OA architecture, what a biologist would call the nervous system Panko terms the new "technostructure." It cuts across all departments ("organs"), and so is organizational, while supporting many different "applistructures" (functions), to be at the same time individual.

"Office automation has traditionally foreseen its products being delivered through workstations built for OA, over OA networks, with OA servers providing OA services," he explains. "We now see, however, that the ideas of workstations, networks, and servers form a general picture of how *all* services will be provided in the future, whether these services fall into the traditional domains of DP, OA, engineering computation, or any other area. In short, we are likely to see a split between the *technostructure* built to provide [basic] services and an *applistructure* used to provide specific services. OA may well survive as a distinct part of the applistructure, but will use a technostructure created for the firm's information delivery needs."

Managing Future Technology

User-group director John Connell takes a similar view of technology's future course. It moves, he says, "in two directions, two integrated directions."

One direction, Connell says, is toward the continuing build-up of large data and word processing centers — again, centers to handle high-volume, Type I transactions. And the other direction is "out into the office," with technology showing up at every workstation, tied together.

"Now, these are two separate thrusts," Connell said in an interview, "and I believe we must come up with organizational approaches that address each separately," managed by two separate groups of people. "Maybe there's someone on top who's in charge overall. Nevertheless, we must recognize that these are separate kinds of activities, but that the architecture of the technology has to support both. You can't have them so divorced that one's going in one direction and the other in another. And yet, the talents required over in the big center are entirely different from the talents required in the office," says Connell.

Regarding how the technology should be managed, Connell suggested that today's answer, of in-house information centers and software libraries, may be only an interim stop on the way to better solutions tomorrow. Too often, he said, these resources are not designed from a user point of view, but from that of the data center.

"The information center manager tells the user, 'Here, you can use my Digital computer.' But that's not what the user wants. The user wants to know, 'How can I use *my* computer and occasionally pull information out of your computer?'"

Connell believes that to properly move out OA technology, a large portion of development work and programming must also move out to the user community. "The central group will design the big frameworks, the architecture, and worry about interfaces between one system and another. But the fleshing out of the system, programming it and so forth, will be done in the user community, by their own local people."

The Corning Example

Connell cites the example of Corning Glass, where MIS abandoned a little-used information center after talking matters over with the people it was supposed to serve. Again, the issue hinged on users not wanting to do things MIS's way but their own way. Here, MIS listened, and soon set up an in-house PC store, inviting users to take the equipment to their workstations. MIS agreed to support them as needed, but otherwise would not get involved in the work. Later, MIS asked users to participate in councils of their own which would collaborate with MIS on matters of information policy.

"But it took a very enlightened MIS director," Connell notes, "to be willing to back down from the traditional way of doing things and say, okay, it's time for a change."

A number of other MIS executives have also shown willingness to deal differently with the user community. "But for others," Connell says ruefully, "control is still their number one desire in life. It's been drilled into them: control, control. I know — I did it myself and so did everyone else. We hogged it all in the data center. And at the time it had some merit, because the only way you could process was with big mainframes. But now with networks you can move it out."

In short, the applistructure/technostructure architectures, by whatever name, which Panko, Connell, and some in the AMS Foundation study group see emerging, will require corresponding management structures, different from what they have been.

The successful DP and MIS managers will continue running their own large centers, but will also facilitate the use of technology by others rather than performing services for them. The successful managers of departments will be sufficiently able to handle their own computer work, with perhaps some MIS help and that of programmers or other "techies" resident in each department.

And the successful future administrative manager could well be that generalist who oversees all information activities in the many-layered, differently typed, yet seamless environment which the integrated office will have become.

Chapter Nine

LEARNING FROM OTHERS — BEFOREHAND

No one implements office automation "cold." Executives responsible for planning and introducing OA have abundant resources at
their disposal to help ensure that their efforts will succeed. And from
AMS Foundation research into the steps they take prior to implementation, it appears most planners do indeed make use of them.

The most widely availed, and perhaps most easily availed, sources
of outside knowledge are conferences and seminars addressing OA or
some aspect of it. Eighty-seven percent of respondents in the study
said they attended such events or sent coworkers prior to OA implementation (see Table 8-1).

The next most prevalent step — visiting other organizations' automated offices — was employed by 60 percent of the group.

So far these methods present no controversy. But when it comes to
the "insurance policy" that scored third in the data — the setting up
of pilot studies — opinions differ within OA circles as to how useful
the step really is. The same can be said for other pre-implementation
moves that were mentioned, like asking a vendor to take a survey or
calling in a consultant.

Test or Model?

What could be wrong with setting up a pilot study to test how the
equipment works and how workers take to it before making any large-
scale commitment to OA? Isn't this a sensible way of testing the

waters without getting "all wet"?

The answer is yes, providing everyone remembers that it is a test and only a test, and that it truly be an objective test.

What too often happens, say critics, is that a pilot project's role as a test is soon forgotten. It becomes a model instead — its good features as well as bad becoming ingrained in future systems. Sometimes the limited-function nature of the pilot, if also carried forward, could actually stunt the growth of later organization-wide or "global" systems which today's technology more readily supports.

"The risk," says a report of The Diebold Group, "is that these pilots, designed as experiments, will pass on their limitations and weaknesses to the companywide systems that subsequently will be implemented." Regardless of planners' intentions, the report implies, these trials, set up for the here-and-now, influence the extension of OA here-and-later.

What is needed, Diebold and other "globalists" say, even before all OA issues are settled or the impacts understood, is a new and larger view: one company, one system, one organic workplace. A test monitored in that kind of holistic planning environment might have value as long as its "proofs" and implications are kept proportionate to the place it occupies in the larger scheme of things.

Stacking the Deck

Another criticism of pilots is that their leaders often play with stacked decks. They choose test sites, it is said, where "success" is virtually assured. Plainly, results will more likely be positive if advanced technology is first tried out in a computer-literate section where people are eager to use it, than in some area where fears must be overcome and heavy training undertaken. But what will "success" in the first instance tell you, critics ask, about the use of OA technology among the office rank-and-file?

Yet a plausible counterargument holds that success breeds success, and that by being able to show off a section in which OA "works" and where workers happily vouch for it, planners can sell the need to change more readily to the skeptics and the fearful.

"Find out what users are rated on. . .what part of their job their pay is based on, and give them a system to help perform their duties," said a vice president of office systems for a New York investment house in a recent magazine supplement devoted to OA. "The application *you bring up first* [emphasis added] sets the tone for everything else to come."

What is this — more pilot maneuvering, more blatant deck stacking, or practical wisdom? Is it perhaps both? Only a goal-oriented OA management can know for sure, and only then if the organizational visions and systems architectures are comprehensive enough.

For OA planners, as for anyone else, success has more allure than failure, even in tests where a failure may have longer-term value. And who can deny that a contagious sense of user enthusiasm is not an asset worth cultivating at any stage of OA development? Pilots, then, if they are to have more than surface value, must be approached by planners as objectively as possible, with full understanding of what one *ought* to get out of them and not just what one *would like* to get out of them.

Pilots must validate OA's worth in typical, pertinent work situations. The ideal test group is emotionally neutral toward OA, willing to try it, and one that planners deem a good organizational place for OA to start.

The Vendor Survey

Office systems surveys taken by vendors eager to sell OA products are another of those temptations that planners know in their hearts have drawbacks — but they consent to them anyway.

At any rate, 44 percent of respondents in the AMS Foundation study said they consented. The temptation, of course, lies in the vendor's offer to conduct the survey gratis or at low cost and then make systems recommendations based on the findings. The drawback, for all the protestations to the contrary, is that the work has commercial bias. It skews, in the aggregate, toward overcapacity — toward recommending more equipment than actually needed — and certainly toward the product line of the survey-taking vendor.

Yet here again, with a bit of watchfulness, OA planners have a source of useful knowledge. The methodologies and formulas in the surveys are tied to some viable base of previous user experience. If nothing else, the results can serve as a low-cost second opinion to place against one's own study, taken in-house or with the help of a consultant. And more than a third of the OA planners in this study did seek consultants' help, and as shown in Chapter Four, many who didn't wished they had.

Consultants, like vendors, are the bane and the blessing of OA implementation. For every planner like the government manager who declared, "Never trust vendors," many others openly praise the help and concern shown by their equipment reps in the long months of getting the system on-line. And for all the well-known tales about consultants — that they borrow your watch to tell you the time, that they offer canned solutions, that they hang on too long after a project is over because they sell new "problems" to management, and that they cost too much — they bring to clients the value of an outside perspective.

Consultants can ask the seemingly obvious, yet often key, questions that no one within the organization thinks (or dares) to ask. They bring broad-scale knowledge and experience formed in work with many clients — assuming they have such track records, and it's good if they do — which again no one in the organization could match.

To put it bluntly, consultants also often serve as a front for actions which management wants to take but does not wish to move ahead on alone, like eliminating a department. So, armed with this hidden agenda, the consultant "builds a case" for that outcome.

Should vendors and consultants be an integral part of the OA planning effort? Again the answer is yes, with the following provisions:

- providing the hiring planner corrects for biases built into the vendor studies, and makes no commitments beforehand to purchase any equipment;
- providing the planner, as client, defines the assistance sought and limits the consultant's contract to particular projects and problems;
- providing, in both cases, that employees are told what the outsiders will be doing in their work areas and that their concerns are dealt with frankly and fairly;
- providing the planners who hire the outsiders keep their minds and eyes wide open at all times.

User Groups

Somewhere between the concentrated dealings with a vendor or consultant and the often remote relationship with a conference speaker lies a mid-course source of OA knowledge: the user group.

Focused yet informal, such groups bring peers together to trade OA experiences and information, often under ground rules of mutual confidentiality. Forty-four percent of the AMS Foundation survey respondents said they belong to user groups. A somewhat higher number, 53 percent, said they sought peer advice apart from user groups prior to implementing OA.

It is not always easy to find user groups, even though many exist. Ad hoc and limited perhaps to one city or region, they often resemble networks of colleagues more than formalized groups or associations. Because company information tends to be traded openly within them, members are often chary about letting in strangers. Surely they would not want OA plans openly gossiped about before their own staffs learned of them. A journalist who has run across such clandestine groups in the course of business coverage said members sometimes were literally aghast at being discovered, "as though planning for OA were some kind of disgrace."

Other user groups, however, operate openly. Perhaps best known is John Connell's Office Technology Research Group, based in Pasadena. Members meet twice a year in unhurried locales like Coronado, California, and Hilton Head, South Carolina. Supported by member fees (currently $2,400 a calendar quarter), ORTG provides library and newsletter services, and permits three representatives of the member organization to attend the semi-annual sessions.

Another group is the Society of Office Automation Professionals (SOAP), started by noted consultant N. Dean Meyer in 1981 as an association of office automation users, consultants and trainers. Now operating out of the AMS Building in Willow Grove, Pennsylvania, SOAP is in the process of forming local interest groups in cities throughout the country. The annual membership fee is a nominal $50 and includes a monthly and quarterly newsletter and a member directory.

Some groups operate under the aegis of consulting firms. The Diebold Group sponsors the Diebold Automation Group. Booz • Allen operates a broad-based Multi-Client Program. Several firms sponsor deep-probing groups of a dozen members or so.

And in a less specific way, because their reasons for being are many, local chapters of professional associations like the Administrative Management Society (AMS), headquartered in Willow Grove, Pennsylvania, often prove excellent vehicles for sharing user experiences.

The literature of OA contains abundant forewarnings on the difficulties of implementing OA successfully. The prudent steps men-

tioned here, of finding out beforehand what to expect and how to
avoid the worst, mitigate — but do not eliminate — the difficulties.
Implementors know there are problems; how many and how bad are
what we analyze next.

Chapter Ten
OA PLANNERS' BIGGEST PROBLEMS

What is the most critical problem in automating the office successfully?

In *The Office Revolution: Strategies for Managing Tomorrow's Workforce,* the first monograph in this set of four, Dr. Harold T. Smith, C.A.M., reported on the wide differences of opinion which exist among various concerned groups when asked that question. Dr. Smith researched four separate groups: (1) OA authorities, (2) administrative managers who belong to AMS, (3) managers involved with OA who belong to the Office Technology Research Group, and (4) white-collar workers, asking each to name the most critical OA problem as they saw it.

Each group complied, and from their answers another critical problem loomed, a problem none of them had mentioned specifically but which was plain enough when their responses were compared: the groups harbored widely different perceptions of what OA meant to them and what the priorities for successful OA should be.

The workers voiced their greatest concern over possible mental and physical effects of OA, and the fear of layoffs. The two groups of managers, whose views on most matters were fairly close, mainly worried about gaining top-management support for OA, and of demonstrating OA's payoff. But the experts — consultants mainly — said "measuring non-repetitive managerial and professional work" was OA's most critical problem, although they agreed with the managers that proving payoff was also a top concern.

"Perhaps there's nothing surprising in any of this," said editorial comment on the research in *Office Administration and Automation*. "Employees naturally view automation as job-threatening, middle managers always worry about the higher-ups, and consultants love to measure work — even when it's a non-repetitive, professional kind of work that's all but impossible to hold up to a yardstick." Still, one could see trouble ahead if these differences were not reconciled.

A Different Outlook

Dr. Smith's research, conducted in 1983, covered a cross-section of North American business organizations — those involved in OA projects as well as those which were not. It thus drew from a wide pool of potential respondents, and the opinions of 362 employees, 216 AMS-member managers, 136 ORTG-member managers, and 16 authorities were actually used in the study.

Research for the present volume drew from a necessarily much smaller pool of respondents, subscribers to the OA newsletter, *Impact: Office Automation,* who are, for the most part, experienced in OA planning and implementation.

While the two studies cannot be directly compared because of statistical and other constraints, it is nonetheless revealing that the opinions of the 45 "veterans" in the study for this book differ significantly from the views of respondents in the earlier study, who represent a less OA-focused audience. Again this need not be surprising: opinions shaped in the light of actual experience will often part company from perceptions shaped by those without direct experience.

In any case, "people problems" — termed *maintaining a human perspective* in the earlier study and the number one critical issue among all groups combined — was not the most-cited problem in the later study.

The biggest management problem encountered by the OA veterans in this latest study has been the *turf battles* which erupted among departments during OA planning and implementation — one-third of the respondents (the largest percentage) rated this as a major problem (see Table 10-1).

The second most vexing issue in this study is technology. Twenty-six percent of respondents in hindsight called *lack of equipment standards* a major problem.

Even so, people problems were plenty big enough. Nearly half of

Table 10-1
The Biggest Problems Encountered
During OA Planning/Implementation

Problem	Percent of Respondents Rating Problem as Major
"Turf" battles (among departments, etc.) ...	33
Lack of equipment standards...............	26
User anxieties toward OA	25
Lack of top management support	21
Inability of planners to agree..............	14
Unexpectedly higher costs of proposed new system	9
Actual user resistance (beyond anxiety) to OA ..	7
Top management interference	—
Union actions or threats..................	—

Note: Percentages are composites of 4 and 5 ratings on a scale of 0 to 5, 0 meaning no problem, 5 meaning big. See Table 10-3 for a finer breakdown of ratings. Thirty-two percent of respondents also named other problems encountered; these are described more fully in text.

the OA planners rated *user anxieties toward OA* as a bigger-than-average issue (three to five) on the 0-to-5 scale (see Table 10-3). But a more extreme form of the problem, *actual user resistance to OA,* was rated above average by only 19 percent of the group. In fact, 60 percent called this a small problem (see Table 10-2).

The fact that worker-related problems are not *the* critical issue in this study may also be a reflection of the kinds of goals this particular group of OA planners pursues, as well as their OA experience level.

As we saw in Chapter Two, reducing personnel was the least of their major objectives. For most, the goals were increased information accuracy and increased speed of delivery.

Also, these goals, as well as other insights, suggest this group has a certain maturity toward OA understanding and usage, an idea supported by the length of time in which many of their organizations have been involved with advanced office systems of some kind. Thus,

Table 10-2
The "Least Bothersome Problems"
of OA Planning/Implementation

Problem	Percent of Respondents Rating Problem as Small
Union action or threats	95
Top management interference	79
Actual user resistance (beyond anxiety) to OA . .	60
Lack of top management support	58
Inability of planners to agree	49
User anxieties toward OA	41
Lack of equipment standards..............	35
Unexpectedly higher cost of proposed new system	29
"Turf" battles (among departments, etc.) ...	24

Note: Percentages are composites of *0* and *1* ratings on a scale of *0* to *5*, *0* meaning no problem, *5* meaning big. Chart is thus an approximate statistical mirror image of Table 10-1. See Table 10-3 for a finer breakdown of ratings.

the data may be read as indicating that, at least for many, staff reduction is *no longer* a high OA goal; it has already happened, and user resistance and anxiety has already been overcome.

Attention then turns to managerial and professional support. In such circumstances, worker anxiety and resistance would not be the major headache.

The existence of turf fights also fits this scenario. As knowledge workers increasingly use and help to set policy concerning personal computers and other equipment (Chapter Eight), and as office automation inexorably changes the ways in which people work (Chapter Two), the issue of "who controls?" becomes threatening to some and contentious for many. It is a people problem, yet, but on an executive level.

Table 10-3

How Office Automators Rate Nine Common Problems

Problems:	Rating:	Small	IMPORTANCE				Big
		0	**1**	**2**	**3**	**4**	**5**
Unexpectedly higher cost of proposed new system		7	22	32	30	9	—
User anxieties toward OA		18	23	14	20	16	9
Actual user resistance to OA		35	25	21	12	5	2
Union actions or threats		88	7	3	2	—	—
Lack of equipment standards		19	16	20	19	20	6
"Turf" battles		14	10	17	26	16	17
Top management interference . . .		48	31	14	7	—	—
Lack of top management support . .		42	16	14	7	14	7
Inability of planners to agree		20	29	27	10	12	2

(% of Respondents Rating)

Note: Figures show percentage of respondents giving particular ratings to various problems. In a fill-in question, 32 percent of respondents cited additional problems, described more fully in text.

The Top Management "Problem"

From what they report, OA planners want — and get — top management support, yet some would draw a line between support and interference.

It may be only a statistical coincidence, but 21 percent of the group cited *lack of top management support* as a major problem, an exact complement of the 79 percent who reported that top management involvement in the OA effort was strong (Table 6-1, Chapter Six).

As noted earlier, top management support sometimes blows hot and cold. Senior executives do not fully understand OA and expect too much. Or, they say they are for it but then do not fund it adequately. Or, they overrule the equipment recommendations of the planning group, naming other "reliable" vendors out of excessive caution or even personal favoritism.

Understandably, what OA planners want from top management is the kind of support that signs the checks and broadcasts pep talks to all users, compelling obedience when the systems changes come — and, of course, a pat on the back, if not more tangible recognition, for the planners themselves. In the eyes of some, anything more amounts to unwarranted interference.

Top management must often walk a fine line in properly supporting OA. But judging from the comments of most planners, for all the difficulties their seniors present, they would rather have them involved than distant and uncaring.

The Standards Dilemma

The problem of systems incompatibility was never mentioned as such in the earlier study by Dr. Smith. Yet in the survey reported in this book nearly half call lack of equipment standards a bigger-than-average OA problem. This is one indication, perhaps, of why half the group opted to start new systems rather than build on what they had. What they had and what they wanted may not have been able to work together.

It is all well and good to say that all OA equipment should be able to interface with everything else. Indeed, many alien brands of merchandise *can* be made to understand each other through the use of modems, translators, and other "black box" devices. That such link-ups are often clumsy, slow, and expensive is another story, but the fact that they exist enables vendors to tell the truth (if not the whole truth) in stating that, yes, their equipment is compatible with all those others.

It is well known by now that many so-called IBM-compatible personal computers are not totally compatible with IBM PCs even through they use the same 16-bit chips, the same operating system, and run the same programs. What is different is the way the various machines format "the same" disks and store "the same" information on them — and that makes all the difference, as many frustrated early buyers discovered after the purchase. But these were, at best, matters of vendor subterfuge which could be overcome with consumer education. What may never be overcome, says OA realists — and maybe never should — is the existence of real, ingrained incompatibilities within OA systems, caused sometimes by calculated competitive maneuvering among vendors, but more often by the unstoppable advance of technology.

At least three major brands of operating systems — CP/M, MS-

DOS, and UNIX — coexist today, and what's to stop better ones from arriving tomorrow? All are basic system-readying programs which enable computers to then run other programs — word processing, spreadsheet, and so on — devoted to specific applications. Efforts go forward to make CP/M-based software more adaptable to MS-DOS-based hardware, and vice versa, and to enable systems of whatever base to communicate more easily.

Still, a world of differentiated systems continues to build, and while MS-DOS has become something of a *de facto* standard because of its use in the IBM PC line, and while work goes on to bridge intersystem gaps, much of this world will still confront the OA manager, 1990 and beyond.

Ethernet, the Xerox-offered coaxial local area network, has also been widely accepted as one LAN standard. But a coexisting environment grows, made up of PBX-based LANs using twisted-pair telephone wires. Again, through "gateways" and other interfaces, the architectures are not out of each other's reach, but these are only two in a multitude of not-always-compatible telecommunications options available (or soon available) to OA implementors.

Yes, this ill-fitting multitude is sometimes vexing, but, OA progressives ask, isn't it really a good thing? Isn't it good that an Apple, for example, dares to confront the *de facto* powers of an IBM and bring some real competition to the industry? Isn't it good that Ethernet never was fixed as the one, now-and-forever LAN standard? Assuming such a move were ever possible, it would have stifled LAN progress. Many already consider Ethernet no longer state-of-the-art.

A central argument against standards — desirable as standards so often seem — is that they thwart technological advances. On the other hand it is sometimes charged that major companies like IBM, powerful enough to set *de facto* standards, give lip service to the idea of negotiated industry standards, but make no meaningful moves toward establishing them because of the competitive advantage they already hold.

The bottom line for OA planners is that lack of equipment standards will continue to be a fact of OA life. And because OA growth will inevitably create more situations in which users seek to integrate dissimilar products, the industry will continue to offer increasingly better, "transparent" solutions. But the problems will continue to cause vexations.

White-Collar Unions

In most popular treatments of life in the automated office, the writer or narrator usually gets around to the question of whether all this new equipment, especially the VDT, poses a health threat, and whether this in turn will lead to a large-scale white-collar union movement.

Judging by the replies of the OA planners queried (and indeed by the sluggish growth of white-collar unions over the last quarter century), the answer, as far as any labor movement is concerned, would have to be no. How many in the survey group said *union actions or threats* were a major problem during OA planning and implementation? None.

This does not mean, however, that unions could not arise. While less than 10 percent of all U.S. clerical workers currently belong to unions, and while white-collar union membership has grown less than one percent per year since 1968, efforts to organize office workers have never been stronger than now.

Watching their blue-collar ranks decline along with the decline of "smokestack" industries generally, union leaders know their future has to lie with the office if they are to have any future at all. Truckers, for example, today account for only 10 percent of members in the Teamsters Union, which has lost 700,000 of these constituents since 1979. Teamster president Jackie Presser has openly declared his intention to recruit white-collar workers. Locals like AFL-CIO 925 concentrate on women office workers while associations like 9to5, both headed by outspoken women's advocate Karen Nussbaum, also raise issues such as equal pay, comparable worth, and the possibility of hazardous emissions from VDTs.

It is possible that the decline in blue-collar ranks has so enfeebled organized labor that it cannot recoup in time, and with the proper skills, to recruit successfully in the office. It is even more probable that management, adept at fostering worker satisfaction through many job-related programs and benefits (not the least of which are the attractive office environments so many employers provide) will successfully introduce OA and still live union-free.

However, the "health issue" could be a sleeper, although VDT vendors steadfastly deny there is any problem of safety. If there is something to these reports of "clusters" of female VDT workers having miscarriages or deformed babies, OA — or at least the VDT, a major tool of OA — could be in serious trouble. At the very least, reports of backache, eyestrain, and high levels of tension have been documented among VDT workers.

According to NIOSH, the National Institute for Occupational Safety and Health, while VDTs do send out minute amounts of electromagnetic field and X-ray radiations, these emissions are extremely low, well below current Federal standards for safe exposure. As outlined in the second part of the AMS Foundation's study, *The Office Environment: Automation's Impact on Tomorrow's Workforce,* by Wilbert Galitz, the number of research studies conducted on the effect of VDTs on office workers has failed to produce any evidence of serious health hazards.

And yet, a disturbing article, "The Mind Fields," dealing with what is known and not known about low-level electromagnetic fields, in the science magazine, *Omni* (February, 1985), tells of seven miscarriages and three cases of congenital defects among 12 pregnant workers operating video terminals at the Defense Logistics Agency in Marietta, Georgia. Four VDT operators in the *Toronto Star's* classified department gave birth to deformed children, the magazine said, while three who did not use VDTs had normal babies. And 12 out of 55 pregnant VDT operators at the Department of Employment in Runcorn, England, were said to have borne malformed babies.

Some of these abnormalities have been explained away as stemming from other prenatal causes, and manufacturers continue to insist that there is simply no evidence to prove that VDTs are hazardous, but as the *Omni* piece points out, there is scant objective research in this field to prove much of anything. NIOSH is said to have additional research under way. It is noteworthy that the Computer and Business Equipment Manufacturers Association (CBEMA), has been chosen to head a coalition of associations banding together to counter misinformation in the press, and to direct the group's lobbying efforts in legislatures where bills restricting VDT use are up for consideration.

Health concerns and office unionization may yet wind up being as much as the non-issues they were in this recent AMS Foundation survey. Still, as pointed out by Wilbert Galitz in *The Office Environment* book of this series, in this area there is much potentially at stake, and prudent OA planners would be well advised to "stay tuned."

The Problem of Too Little Space

One problem too often ignored in OA planning, yet painfully encountered during implementation, is where to put the pieces. Computers, disk drives, modems, printers, and even user manuals are bulky physical objects that eat space on work surfaces and throughout the work environment — to say nothing of all the paper which OA does not eliminate. Everything looks so neat in systems diagrams. However, in the traditional office workplace, now suddenly automated, it can look a mess.

Cabling presents an especially challenging problem, particularly in open-plan or "landscaped" environments where the ease and economy of rearranging workstation layouts has always been a strong point.

Members of councils like the Office Landscape Users Group and the Organization of Facilities Managers and Planners cite savings of up to 90 percent in rearranging open-plan layouts as opposed to remodeling space having private offices and fixed walls. That was, however, before networks of cables for computers, word processors, and other automated devices came upon the scene. Now, the need to snake wires into spaces under raised floors or above dropped ceilings, or through the raceways of panels or "systems furniture," is robbing the office landscape of much of the flexibility it once enjoyed, and causes office administrators to think twice before going ahead with layout changes.

Yet, as Wilbert Galitz pointed out so emphatically in *The Office Environment,* change is a constant in today's and tomorrow's workplace. The impact of OA on the ergonomic and environmental aspects of the workplace is substantial, and must be planned for in advance of an OA implementation effort.

Other Problems

What else has gone wrong in the office automation planning process? Other comments volunteered by some of the planners include the following:

- "Not enough money. Not enough planning personnel. And management was unable to perceive the value-added benefits."
- "The vendor was unable to deliver the hardware; then 'technological interruptions.' Meanwhile, a faster than expected growth in user demand."

- "Existing positions were reclassified. Also, our biennial budget cycle made it difficult to identify [and cost out] state-of-the-art technology three years away."
- "Central DP didn't want to have PCs entering the system" — another turf fight?
- "Equipment incompatibility" — standards again.

The OA world looks different to the seasoned traveler who has seen it up close than to the more-or-less interested spectator who views it from afar. Some of the most revealing findings of the study reported in this book come not from the tables of data derived from their replies, useful as they are, but from the trenchant comments on what they would do differently — or advise others to do and not do — in planning for OA anew.

In the next chapter we will hear these veterans speak.

Chapter Eleven
ADVICE FROM OA VETERANS

Throughout this book we have heard OA executives "speak" on the goals, problems, and systems architectures of office automation, and of their involvements with top management during the planning process. Their unvarnished views of "the way it is" play effective counterpoint to the reams of planning theory found elsewhere in office automation literature.

In surveying these executives, the AMS Foundation encouraged them to be frank, first by promising them personal anonymity if they wished it, and second by probing insistently into the heart of planning reality by asking essentially the same question several different ways:

- What changes in systems design would you make were you to begin your OA effort anew?
- What changes in planning, timing, and/or budgeting would you make were you to begin your OA project anew?
- Based on your OA experiences, what one or two things overall would you do differently were you to start the project again?
- What basic advice would you give to fellow managers who are about to begin the process of planning for and implementing OA?

Selected replies, as noted, have already been quoted to amplify points in the course of this work. However, to fully grasp the range and challenge which OA planning involves, one must assemble the veterans and hear, one by one, their experiences as they relate them.

Here, then, in no special order except to follow the question sequence above, are their comments, lightly edited where needed for brevity and clarity.

Changes in Systems Design?

I (we) would:
- Probably look at different modes because the nature of the business has changed.
- Obtain more information on software and also more knowledge of networking.
- *Test in depth* any software considerations before purchase. I would not consider hardware which could not run major software, such as Symphony, dBase III, WordStar.
- Have less volunteer assistance and more paid consultants.
- Go IBM, use 3270 emulation terminals, and bring them to managers' desks.
- Distribute the OA function to Wang, and use IBM as a network controller and distribution facility.
- Begin with a central organization to administer OA. I'd provide more resources to support the effort and allow more time for orientation and training.
- Start planning earlier.
- We considered WP in depth and guessed how it could fit into OA. I would have preferred to look at company communication needs first, map out strategy for OA, and then implement WP as a first step.
- Changes? None really. The plan was informal and then formalized. It's now working.

Changes in Planning, Timing, Budgeting?

I (we) would:
- Increase the number of people on the planning committee.
- At the beginning, ensure corporate financial commitment to the plan.
- Work to maintain top management support while trying to involve users more closely.
- Try to get the total organization thinking about OA.
- Plan and budget for more programming than originally anticipated.

- Get a commitment in writing as to what was actually required from management.
- Shorten the time between when we are ready to implement and when funds would be available.
- Plan along project life-cycle functions to assure that applications of significant value drive solutions. Plan short- and long-term to have full-time team members and part-time user participants develop standalone and mainframe access over a three- to five-year span. Use more objective and scientific methodology and evaluation, avoiding impressionistic answers.
- Take more time.
- Start participatory planning earlier.
- Gain high-level management sponsorship for the OA study, and use that sponsorship to gain user participation. Then use the study to map out long-range requirements and strategy.
- Nothing major — possibly increased user-manager coordination.

Do Differently Overall?

I (we) would:
- Start sooner.
- Not allow top management to override user preferences as to hardware and software suppliers.
- Plan more carefully. Experiment with equipment. Visit vendor plants.
- Establish a clearer line of authority and responsibility.
- Enlist greater management support up front.
- Qualify management requirements. Include more users in decision making.
- Provide more resources and people to support a pilot program.
- Implement accounting applications first, rather than database management. Quicker results at less cost would have pleased our board more.
- Have experienced people on site during planning and implementation stages to oversee actual progress and to assist in eliminating problems.
- Assign one generalist, with a broad view of organizational goals, to mediate among DP, WP, and telecommunications.
- Find out more about micro networks, speeds, numbers of users, degredation of systems, and so on.

- Emphasize pilots with communications requirements.
- Obtain stronger commitments from early end users so that we could proceed directly to other end users sooner; hire more and better experienced operations personnel sooner in the process.
- Establish and document measurable goals.
- Check out software problems better than we did, and see to available training for equipment as well as for software.
- Investigate all possible vendors before making decision.
- Send employees to seminars to get them interested before implementation.
- Make higher cost estimates. Try to obtain more systematic user input.
- Visit other organizations which have the equipment already in use. This would enable us to determine actual space needed for equipment and assess noise levels while the equipment is working. Also, we'd talk to operators using the equipment.
- Get top management's commitment to financial support and standardization.
- Attempt to be more aware of how we'd fund the payments.
- What do you mean "start the project again"? It's not a project — it's a *process*! Dynamic. That's how I view it and how I "work" it.

Advice to Fellow Managers?

- Keep it simple. Keep as many options open as possible. Don't allow any vendor to lock you into a product line.
- Obtain a complete go-ahead from top management so they can't sit back and say, "I told you so."
- Don't underestimate cost or personnel needs — personnel for planning and personnel to manage the implementation.
- Check on the training available — where, when, how long, how many people will be trained?
- Become actively involved with all phases of OA. Allow employees working on equipment ample time for training. Keep the training on your site whenever possible.
- Be sure to have the right equipment for your objectives. Don't forget to properly plan the layout.
- Cultivate end users; see if they'll accept some failures in trade for getting systems out to them early. Realize that the mechanics of OA (cable runs, schedules, etc.) are more complicated and trouble-

some than conceptual planning and implementation. Never trust vendors.

- Begin small and be successful.
- Demand implementation of one function at a time and give realistic schedules to management.
- Look at microcomputer solutions. Their speed, memory, and software make them cheaper, viable solutions.
- Make sure pilots address real business problems. Make sure systems reach "critical mass" quickly.
- Obtain top management commitment. Use a pilot-project approach. Obtain top-level personnel.
- Keep it simple. Evaluate the needs of your users based on their own input. Don't let overwhelming "options" sway you from your planned course in meeting those needs.
- Put together your strategic plan first — the purpose of your organization and its goals. Then ask how OA can meet those goals in all its forms.
- Know present systems thoroughly. Do cost-saving analyses. Visit a user site and get their feedback. Document your plans and schedules. Make sure your (planning) chairperson is experienced or at least has been thoroughly briefed.
- Find your software first, then choose hardware. Look hard for software that allows users to do their own programming rather than depending on conventional languages that require a programmer or consultant.
- Become knowledgeable in the OA area. Find a vendor you can trust. Include as many users as you can in the planning stage. Research all costs — equipment, maintenance, supplies.
- Know that OA is a complex undertaking and requires a balance between technical and human factors if the automation is to really work in a day-to-day environment.
- Don't underestimate training and support costs. Don't overestimate savings and paper reduction. Don't wait for the next technological development.
- Operate on both of two levels: One — get top management support by identifying quick-payback, obvious-benefit projects to get early credibility. Two — use a committee with MIS commitment to define long-range issues and clarify the OA turf.
- Know your organization. Where and what are its fears? Realistically evaluate your OA limits in time and expenditure. Ask: what

specifically do you want OA to accomplish? Give yourself plenty of time to stop and evaluate your progress and direction.

- Try a pilot operation. Stress OA orientation and training to users. Provide adequate support resources. Select a "mainstream" vendor who will be around for the long haul.
- Study more up front. Keep scope wide. Phase in implementation.
- Obtain senior management commitment. Determine methodology and justification rules in advance. Formally organize and fund user involvement and commitment. Clarify technology.
- Go slow. Develop a clear understanding of objectives. Do not be blinded by technological frills. Establish standards for equipment and furniture. Develop standards for measuring effectiveness.
- Do it in a participative way early so you have support when you get to the important communications issues.
- Start slowly.
- Ask yourself three questions with respect to all equipment: (1) Do you need it? (2) Who will operate it? (3) Will those people understand it easily?
- Stay flexible; be aware of networking.
- Use independent, reputable, outside paid consultants. Have one senior staff person who is knowledgeable in the company's operations serve as liaison. Identify, organize, and prioritize current manual procedures and systems before automating.
- Make sure all manufacturers and suppliers are supported locally for quick service. Use only those that meet your requirements at least 95 percent overall, because we all have want lists that *can* be fulfilled, but at higher cost. Make sure your software is not 3,000 miles and three time zones away.
- Look at the big picture and plan long term. Try to achieve vendor uniformity for maximum compatibility of equipment. Clearly establish who's in charge. Make sure you have adequate support.
- Start planning now. This is the problem-prevention phase. Sell executive management early. It will not happen without them.

Chapter 12

THE BOTTOM LINE

"We planned thoroughly and have had few surprises."

"We are very satisfied thus far with our project and planning."

"In all, we spent over two years [planning and implementing] and have emerged with the most sophisticated and most advanced system in our industry."

Regardless of how long, complex, aggravating, and expensive the process may be for planning and implementing office automation, and however imprecise certain benefits, a strong current of satisfaction for having made the effort does gladden the OA community.

As John Diebold points out (Chapter Three), managers using advanced information systems and using also common sense can *know* when they are gaining productive advantage even when they cannot measure those gains precisely. There is an art as well as a science to management, and intuitively sensing when OA investments are in one's business interests is part of that rarified art.

But tangible gains from OA do reach the bottom line, too, and examples of them have been offered throughout this book. A further sampling:

- An electronic filing system at AT&T Technologies has eliminated 500 square feet of file space, allowed a 20 percent reduction of accounts payable clerks, reduced filing backlog by 95 percent and virtually eliminated misfiles. Bottom line for this Kearney, New Jersey, facility: savings of $62,000 a year.

- An audio teleconferencing system, part of an evolving OA architecture at Sun Co., Radnor, Pennsylvania, has alone saved more than $100,000 per year in travel costs. Office automation at the big oil company also includes WP, electronic mail, and a computerized library.
- A freeze-frame videoconferencing system at Health Care Financing Administration, the agency in charge of Medicaid and Medicare, has in a year and a half returned $2.50 in saved travel costs for every dollar spent on the system. In those 18 months, the system handled some 1,000 teleconferences involving 9,000 attendees. This cost $141,824, the agency said; savings and productivity gains totaled $379,852.

In the more confined and measurable world of word processing, productivity gains of 50 and 100 percent over traditional typing operations are today commonplace.

OA's Higher Purpose

At this point, several points must be emphasized. Examples like the above almost of necessity reflect the variety that exists in office work. Yet almost of necessity they focus on only one facet of OA, such as a conferencing system or a filing system. These are components discrete enough to be cost-and-benefit analyzed. The larger OA systems in which they reside, often years in the making, can have grown beyond the bounds where overall measurements may be taken. Too much time has elapsed, too many intangibles have crept in, and early studies may no longer have relevance.

But, again, savings are themselves only one aspect of the OA rationale. The higher purpose of OA, which so often gets lost amid hills of counted beans, is to enable workers to perform better, not merely produce cheaper. Its real value lies in enabling organizations to compete effectively. These, of course, are "soft" values for perceptive managers to intuitively understand, not mechanically calculate.

However, support for calculating OA's "soft" benefits may be on the way. N. Dean Meyer has recently announced plans to conduct research into OA's value-added benefits — those specifically defined as "over and above efficiency" — in applications where OA has become a strategic business resource. He will attempt to use a methodology "that measures benefits even when there are no tangible cost savings." He says his research will help establish the importance of

OA in the minds of executives; stimulate implementors and users to think creatively about high-payoff applications; and assist planners in justifying and evaluating management applications.

A Collaborative Journey

The bottom line for any OA planner, implementor, and user is ultimately meaningful only in light of the goals they set for themselves, their departments, and their organizations.

Office automation is a strategic, collaborative journey. It has as many destinations and reasons for wanting to arrive at them as do travelers at a crowded airport. The comparison is more than apt. The comfort and safety of those flights depend on the skills and talents of many parties — from the on-board crew to the on-ground mechanics to the tower controllers — and on the technology in which they place their trust. The efficacy of the automated office for any particular organization relies on efforts no less professional, interdependent, and well managed.

Asked to sum up what it takes to implement OA successfully, Samantha Simonenko, assistant manager with an advanced office systems group at Michigan Bell in Detroit, Michigan, who was one of many OA planners contacted for this book, thought a moment, then replied:

"Have total commitment. If your company doesn't have it, *you* have it anyway. . . . Document heavily. It reduces misunderstandings. . . . Do your homework. Be prepared for questions in groups. . . . Maintain a sense of humor. Things will go wrong. So it's better to laugh than to cry."

May your own OA journey go smoothly, and your destination, when you arrive, be all that you had hoped for.

Appendix A
DESCRIPTION OF SURVEY

The survey data cited throughout this book was derived from questionnaires sent by the AMS Foundation in the fall of 1984 to a sample of 482 subscribers to *Impact: Office Automation* newsletter, which is published for office automation executives by the Administrative Management Society, an international association for management development. Completed questionnaires were received from 57 respondents, a 12 percent return. Of these, 12 said they were not presently planning for or implementing OA, and were eliminated from survey tabulations.

The remaining base of 45 may seem small by some statistical standards, but comprises a rare concentration of OA knowledge, given today's relative scarcity of developed OA operations. Nowhere in this monograph does the data purport to be more than what it is: the enumerated and proportioned experiences of a specific group of executives who have planned for and implemented OA, or who are involved in that process.

A copy of the survey questionnaire may be obtained by writing to the AMS Foundation, 2360 Maryland Rd., Willow Grove, PA 19090; 215-659-4300.

WHO DID THE PLANNING?

Listed below are the titles of individuals or members of committees responsible for office automation in 45 organizations where OA has been implemented or soon will be implemented, as indicated by respondents to the AMS Foundation survey (see Appendix A). Fifteen of the situations involve one-person undertakings; 30 involve planning committees.

One Person:
1. Coordinator, School of Business Services
2. Comptroller
3. President/CEO
4. Assistant Superintendent (plus "lay committee")
5. Manager, Administrative Services
6. Assistant to President (initially)
7. Manager, Personnel Services
8. Manager, User Support Services
9. Director
10. Office Manager
11. Director of the Department
12. (name, not title)
13. Programmer Analyst
14. General Manager
15. Manager, Information Center

Committee:

16. VP Finance & Administration*
    VP Quality Assurance & Human Resources
    VP Customer Services & Facilities
17. VP & Treasurer
    Director, MIS
    Controller
    Manager, Administrative Services
18. VP Administration*
    Executive VP Finance
    Manager DP
19. Volunteer advisory committee
    Independent systems consulting company
20. MIS Applications Development Manager*
    MIS Senior Systems Analyst
    Corporate Word Processing Supervisor
21. Administrative Manager
    Systems Analyst
22. OA task force with numerous members from most divisions
    Group Manager, Office & Telecommunications Systems &
      Services*
23. Consultant
    General Manager, Administration
    MIS Department Manager
    Director, Customer Service & Data Processing
24. Technology steering committee
    Director of Data Processing
    Director of Administrative Services
    Assistant Director of Data Processing
    Assistant Director of Administrative Services
    Technology Research Specialist
25. Assistant General Manager, Administrative Services*
    Data Processing Manager (later redesignated Director, Office
      Automation Services)
26. Manager, Sales Management Systems*
    Manager, Mechanical Systems & Procedures
27. Manager of Methods & Procedures*
    Senior Systems Analyst (three)

Committee chair

28. Controller*
    Assistant Director for Client Services
    Assistant Director for Special Services
    Assistant Director for Residential Services
    Executive Director
29. VP Research & Planning, MIS*
    Director, Office Systems, Corporate Administration
    VP Corporate Administration
    Director, Systems & Programming, MIS
    Director, Telecommunications, Corporate Administration
    (Composition of committee changed over time)
30. Administrator*
    Quality Assurance Coordinator
31. Secretary-Treasurer*
    Coordinator, Information Systems
    Accounting Supervisor
    Office Manager
    General Superintendent
    Marketing Manager
32. Program Manager, Administrative Systems*
    Management Analyst, Administrative Systems
    Systems Analyst, Data Processing
    Communications Analyst, Communication Plans
33. VP Finance*
    Office Manager
    Word Processing Manager
34. (Checked, but no titles)
35. National Data Processing Manager
    Industrial Engineering Department study group
    Office Manager, Calgary Plant
36. Not a formal committee but interested parties including:
    Director of MIS
    Supervisor of Word Processing
    Associate Administrator responsible for telecommunications,
       WP
37. Secretary of the Senate*
    Special Assistant to Secretary of the Senate
    Commissioner of Legislative Bill Drafting Commission
    Computer Systems Consultant to Secretary of the Senate
    Director of Operations and Maintenance

Office Automation Project Manager
Office Automation User Services Manager
Office Automation Computer Center Manager

38. VP Productivity Management Department*
    Administrative VP, Commercial Banking Systems
    Third-party consultant

39. VP of Administration
    Administrative Assistant to Division Director (three)
    Systems Analyst (three)
    Controller
    Director of Research

40. Assistant Finance Director
    Records Manager
    New department was organized to implement project: Assistant
      Finance Director became Director of Information & Commu-
      nications Services. Records Manager became Manager of Inte-
      grated Office Systems.

41. Project Leader*
    Analyst (two)
    Trainer
    Administrative Specialist

42. Chairman of the Board
    VP Administration
    President

43. Consultant study
    Input from Management Services staff
    Primary staff person: Coordinator, Office Systems
    Assistant Superintendent, Management Services*

44. Corporate Specialist, Office Automation
    Manager, Communications
    Director of Information Systems

45. Administrative committee

ONE ORGANIZATION'S OA PLANNING CHARTER

The following is excerpted from documents of a major midwestern utility, to suggest the range and depth of office automation planning groups and procedures. Its purpose is illustrative, and not meant as documentation which others should necessarily duplicate.

Interoffice Memo:

Advanced Office Systems (AOS) refers to the process of bringing together the power of telecommunications and the computer, to support the management of the business. Our recently adopted charter of AOS in [company] is attached. It provides that [our] Information Systems Organization will establish and chair an interdepartmental AOS steering committee. We are preparing to form that committee and would like the names of one or more representatives from your organization.

Responsibilities of the committee members include, to:

- Speak with knowledge and authority for the organizations they represent.
- Provide information regarding user needs for AOS.
- Assist in developing corporate solutions to user needs.
- Obtain departmental approval and funding for AOS projects.
- Monitor AOS implementation within their organizations.

  [Name] will chair the committee. Would you please provide [him] with the name of your representative(s) by [date].

Attachment:
Advanced Office Systems

Advanced Office Systems is a term which refers to the application of communications and computing technologies in direct support of managers and associated clerical personnel.

Objective

It is the objective of [company] to improve office productivity through the orderly introduction of AOS. The authority and responsibility for achieving this objective will be Information Systems Organization (ISO). It will optimize integration of AOS in [company] and avoid piecemeal development of short- and long-range plans.

Mission Statement

[Company] will utilize current and future office information and support systems which improve the effectiveness of personnel in their accomplishment of corporate objectives. Specifically, we will use AOS to:

- Improve the effectiveness of individuals and groups in accomplishing their objectives; also to increase productivity of managerial and support personnel through the introduction of new job tools, office equipment, and communication systems.
- Reduce the cost of internal information handling and decision support systems.
- Minimize the duplication of information storage and access locations, at the same time reducing amount of space required for storage of files.
- Create a system of intercommunication that provides timely, accurate transmission of ideas and information between individuals and groups.
- Improve QWL for employees by reducing repetitive and undesirable tasks.

Implementation

ISO will establish and chair an interdepartmental steering committee to:

- Implement AOS.

- Evaluate state-of-the-art development in AOS technology.
- Develop and maintain conceptual models of integrated advanced office systems as a planning tool.
- Reduce duplication of effort currently existing in prototype development.
- Provide advice, counsel, and direction to AOS.
- Develop and employ needs analysis and cost analysis procedures.
- Establish migration plans from existing systems to future advanced office systems.
- Develop measurements which specify effective service tools.
- Increase management awareness of the need for coordinated office solutions.

Interoffice Memo [11 months later]:

An interdepartmental Advanced Office Systems steering group has been formed to provide direction and control for the Corporation in the acquisition, introduction, and implementation of AOS. Procedures have also been developed to support the steering group in carrying out their mission. The attachment outlines the procedures to be followed for AOS requests.

We would appreciate it if you would inform your management team about the AOS steering group and the supporting procedures. If you, or your managers, have any questions about AOS, call your segment coordinator, [name], or [name].

Attachment:

Company-Wide Procedures for Advanced Office Systems Acquisition and Implementation

1. Purpose
Support Services and the Information Systems Organization jointly chair an interdepartmental steering committee named the Advanced Office Systems Steering Group (AOSSG). The AOSSG represents the segments and interfaces with the various users. They obtain departmental concurrence and approvals on AOS service requests and present the status and needs of AOS within their segments to the com-

mittee. The Advanced Office System Coordinating Team (AOSCT) is a sub-group of the AOSSG.

The AOSCT has been established to control the acquisition, introduction, and implementation of AOS for all organizations within [company].

The purpose of this procedure is to establish the AOSSG as the group which provides centralized direction for the Corporation in AOS and to establish AOSCT as the group that will review and concur in all future AOS and word processing system installations. In these areas, this procedure supersedes all previous policy and procedures covering AOS and WP topics.

AOSCT will also act to coordinate its activities with those of other departmental organizations to insure appropriate review of all AOS-related proposals.

2. Definition

Advanced Office Systems (AOS) is a broad term which includes most of the computing and communications technologies that directly support managerial and administrative personnel in the office. It positions the manager as the focal point of office systems and includes, but is not limited to:

- Electronic document and message preparation, including all WP technologies, which record information in magnetic or electrical form. Also included is the input of documents into a text processor by optical character recognition (OCR) methods.
- Electronic document and message communications including interface to the corporate data network and dial-up facilities. This includes all applications which may be defined as "electronic mail."
- Electronic managerial productivity aids including reminder systems, calendars, project and time scheduling, and calculator capability.
- Electronic document storage, access, and output including necessary file access to other information or time-sharing systems through AOS communications functions. This includes all communicating document output devices such as printers and intelligent copiers.
- Other mechanized or automated applications that utilize computer and communications technology to improve managerial productivity or administrative support in the office environment.
- Electronic information processing of local AOS text, records, files,

or database items for direct office or management use. That is, database management, query functions, and report generation associated with "personal" or localized data resources for management functions.

3. AOSCT Organization

The purpose of AOSCT is to provide centralized expertise and resources to review and concur in the development, trials, testing, procedures, acquisition, and implementation of AOS products and services. The AOSCT is jointly chaired by the Support Services Department and ISO and has representatives from the following organizations: [list follows].

4. AOSCT Activities

The AOSCT will prepare and maintain interrelated AOS development activities and will produce AOS guidelines and procedures. The following represent primary functions to be performed in meeting its charter:
- Develop a unified company direction for AOS and publicize that direction through a Fundamental Plan.
- Develop communications and interface standards and guidelines to insure AOS compatibility.
- Establish guidelines for AOS equipment selection.
- Prepare effective study procedures to identify AOS applications.
- Review and process all AOS service requests.
- Study and test OA services and features for possible company use.
- Coordinate necessary documentation and training materials for AOS implementation and administration.
- Develop and maintain measurement standards and monitor the effective performance and adequacy of AOS applications.

5. AOS Service Requests

The following broad guidelines will serve to control all requests for AOS service:
- All requests should be submitted through their segment Steering Group (AOSSG) representative to AOSCT.
- The proper documentation for requests [is explained].
- The AOSCT will insure review of user, technical, and operational considerations and provide a recommendation to the requesting district level manager. This recommendation will include direc-

tions on proceeding with the request if desired/appropriate.

6. *Summary*

The establishment of AOSSG and AOSCT brings together on an interdepartmental basis the necessary expertise to effectively introduce advanced technology in direct support of management personnel. These groups will act to formulate a companywide AOS strategy, communicate that strategy to all departments, and insure compliance with the strategy through review of, and concurrence in, all requests for AOS services.

DATE DUE